THE FURROW AND US

Essays on Soil and Sentiment

Walter Thomas Jack

Edited by
Zachary Michael Jack

University Press of America,® Inc.
Lanham · Boulder · New York · Toronto · Oxford

4501 Forbes Boulevard
Suite 200
Lanham, Maryland 20706
UPA Acquisitions Department (301) 459-3366

PO Box 317
Oxford
OX2 9RU, UK

British Library Cataloging in Publication Information Available

Library of Congress Control Number: 2005929381
ISBN 978-0-7618-3317-8 (paperback : alk. ppr.)

THE FURROW AND US

Contents

Editor's Preface

It has been a distinct pleasure to prepare an edited, updated version of Walter Thomas Jack's conservation classic, *The Furrow and Us*. Nearly sixty years after their original publication, Jack's words, grounded at conception, have aged well.

Few changes proved requisite. When necessary and in keeping with contemporary style, more inclusive language has been introduced; elsewhere the volume has been edited for concision, allowing Jack's persuasive rhetoric to build a body of evidence, free from interruption, in favor of the moldboard plow. The existing scheme of twenty chapters, reflecting Jack's evolving and somewhat improvisational research and writing from approximately 1943 to 1946, benefits from a new tripartite division—"Sentiment," "Soil," and, "Soil and Sentiment"—as well as a reduced chapter count.

In Part One, "Sentiment," Jack offers an intimate account of his first twenty-five years of farming and provides a brief review of agriculture from Ancient Egypt to World War II America. In Part Two, "Soil," the author gives practical advice for maintaining productive soils and raising healthy crops and livestock. Part Three, "Soil and Sentiment," weaves the poetic with the practical, embodying the cross pollination of art and science Jack deems essential for the progressive steward.

This new edition makes possible the inclusion of selected entries from Walter Thomas Jack's farm journals, unearthed in the summer of 2004 for the first time since his untimely death in 1965. The journals reflect Jack's passion for imaginative writing as well as the transcendent, literary quality of his agrarian experience.

Zachary Michael Jack
June 2005
Oswego, Illinois

Acknowledgments

This volume would not have been possible without the support of several entities committed to its realization, most especially North Central College, Foxbriar Farm, and New Buffalo Press. Thank you.

Special thanks to Michael, Edward, and Julia Jack for their unfailing belief in the legacy of Walter Jack and for their preservation of the soil and sentiment he held dear.

The original edition of this volume was copyrighted in 1946 by Dorrance and Company, Inc. of Philadelphia and dedicated "To Helen." Acknowledgments were made "for permission to use certain material" to J. B. Lippincott Co. on behalf of W. W. Weir, author of *Soil Science* and C. A. Price, publisher of *Your Farm* magazine, whom the author thanked for his "valuable material and kindly advice."

This edition is dedicated, naturally, to Walter Thomas Jack, whose literate, thoughtful, and reverent stewardship of the land still serves.

Introduction

THE MAN BEHIND THE FURROW

The time is the 1940s; the place is America, where the modern business of farming is being born. Average farm incomes jump more than 150 percent; average yields double for many grains. The American farmer enjoys a clout never before dreamed of, at home and abroad. Tractors, over two million of them according to agriculture historian David Danbom, transform the landscape, allowing for bigger acreages and more efficient production.[1] In the agricultural universities, research divisions, flush with government money, pioneer herbicides, insecticides, and growth hormones. In the fields, de facto government agents—legions of extension workers and Farm Bureau representatives—canvas rural America, disseminating information about the latest soil science and propagandizing the war and its necessary food inputs.

In the midst of this production frenzy, an Iowa farmer and former Quaker schoolteacher, Walter Thomas Jack, publishes a book with Dorrance and Company in faraway Philadelphia, a book that calls for something revolutionary in its simplicity: sustainability. In 158 plainspoken pages, Jack, dubbing himself an Iowa "dirt farmer," argues what would become key planks in the environmentalist platform some thirty years later, namely that: 1) chemical-intensive farming is a net drain on the environment 2) farmers, not industry scientists or university extension agents, know best how to steward their land, and 3) mother earth only provides if she is provided for, and provision is best made with a simple furrow and plow. Jack's opening paragraphs served as a shot across the bow of every production-mongering scientist, commodities broker, and farmer then pursuing volume at any cost: "To gear our nature to our modern civilization," Jack warned, "is but courting madness."

In the sixty years since the publication of his groundbreaking book of agricultural essays, *The Furrow and Us*, the name Walter Thomas Jack has traveled well beyond its home whistlestop: Cedar County, Iowa. A Quaker by birth and by temperament, Jack's expertise with and passion for agronomy, botany, and ecology is exceeded only by his modesty, a legacy that has understated the national impact of *The Furrow and Us*, which is today included in the premier collection of agriculture works in the United States, the Core Literature of Agriculture housed at Cornell University.

Despite the book's adoption by agriculture, *The Furrow and Us* is a proto-environmentalist piece, anticipating with uncanny accuracy the sustainable and organic farm movements. Of course the reprinting of Jack's book, offered here in its sixtieth anniversary edition, makes content its agricultural devotees, who have heretofore to search the shelves of used and rare bookstores to find, as Dorrance put it, this "dramatic report of a dirt farmer" and "thesis on soil and plant behavior that will serve as a helpful reference." In 1946, the year of its publication, no one, not even the forward-looking editors at Dorrance, fathomed that the themes of Jack's book—sustainable growth, local ecologies, and organic methods—would make his modest manual a sought-after commodity well into its sixtieth year.

In backing the book, which openly disagreed with then agriculture darling Edward H. Faulkner and his contrarian sensation *Plowman's Folly*, Dorrance, a small but productive publishing house based in Philadelphia with tastes for memoir and autobiographical accounts from middle America, took a mighty chance. Three years earlier, Grosset and Dunlap had published *Plowman's Folly* and promoted it aggressively as a book to be read by all regarding the alleged environmental dangers and historic inefficiencies of the moldboard plow and its deep-cutting furrow. Louis Bromfield, author of *Malabar Farm*, famously claimed in *The Reader's Digest*, "Probably no book on an agricultural subject has ever prompted so much discussion in this country." *Time* magazine hardly exaggerated when it called the controversy between till and no-till, the two poles represented by Jack and Faulkner respectively, the "hottest farming argument since the tractor first challenged the horse."

Curiously, the two men fighting this battle—one backed by a huge New York publisher mobilized to capitalize on the controversy, the other an independent farmer working with a small press in order to right a perceived wrong—were themselves both practicing farmers and teachers, not scholars. Faulkner, though college-trained at Cumberland College (then Williamsburg Baptist Institute) and the University of Kentucky, was at the time farming a mere "garden plot of farm scale" near far-flung Elyria, Ohio, and referred to himself on his book's dust jacket as a "soil and crop investigator in private employment." Jack, who offered Iowa "dirt farmer" as his calling card, reveled in his underdog status, caricaturing Faulkner as a hopelessly deluded, perfectly pompous academic. Jack writes, "It is amusing to hear the wild claims of some who stick their thumbs in their vest and fondly imagine they have solved the big riddle." In this middlewestern showdown, Jack had Faulkner painted as the bespectacled vest wearer with a whiff of the city in him, while he, Jack, preferred bib overalls and farm-grounded truths.

In the farming communities of 1940s America—towns whose flagging spirits were then being reawakened by the return of sons and daughters from the war—the controversy caught on, and folks chose up sides. Ron Jager, author of *Eighty Acres* and *The Fate of Family Farming*, remembers the Jack/Faulkner grudge match as a topic of heated conversation in his rural community in Michigan in the late 40s. Harriet Barlow, director of the environmental and literary not-for-profit, The Blue Mountain Center, recalls the Jack/Faulkner hullabaloo was still alive and well in the days of her graduate education in environmental studies in the 1960s and '70s.

In the middle 1940s, the fight over the moldboard plow and its advocates (Jack et. al), and backers of so-called "subsurface tillage," (Faulkner et. al) had just begun, while the fight in Europe and the Pacific was newly ended. Still, the battle underway for the "hearts and minds" of rural America, to borrow a phrase from our contemporary debate, had all the high rhetoric and conspicuous flag waving of a political campaign. Jack, writing in 1946 as the region's veterans returned home to take up the plow again, cautioned against following Faulkner's shortcut of no-tillage and nitrogen fertilizer. "This controversy," he wrote, "must not be left as an enigma for our returning service men, and them I will warn—there is no shortcut in soil building." For southern farmers in particular—many of whom were coming back to small farms devastated by the boll weevil and largely unaided by New Deal initiatives geared towards large scale, commercial farming—Jack offered a traditional, conservation-minded agriculture that would brighten the "dull picture of the plight of the sharecropper" whose prospects could be "altered to one of beauty and plenty if certain natural laws were practiced" in lieu of the application of chemical fertilizers. In the South, in particular, where the clay-ridden, thin soils and perennial poverty made impractical Faulkner's costly, nitrogen-dependent methods, Jack's argument found particular resonance.

Elsewhere, Jack, making hay of the rapidly changing international, political, and cultural scene since Faulkner's publication in 1943, became even more patriotic, taking offense in his Author's Preface to Faulkner, represented therein as "the most ardent critic of the furrow" who, Jack claimed, "ridiculed" the American farmer for not producing yields comparable to the Egyptians, who still used primitive agricultural tools. Jack bristled at Faulkner's anti-democratic insinuation, as did many of the common sense farmers of his era, who saw in Jack a true representative of their yeoman concerns, their respect for natural laws, and their legendary irrepressibility. Jack engages in some populist pulpit-beating in an attempt to answer his fellow farmers' wounded pride, writing, with a patriotism, if not jingoism, characteristic of the day: "No, the American farmer does not envy the crooked stick farmers of any country for they know the furrow is mightier than the Nile."

The battle for the naturalness and the sanctity of the precious topsoil of the American Plains was a battle, in Jack's estimation, for civilization and the maintenance of a leisure class, a class to which Walter Jack, as a farmer, did not belong, but which he nonetheless fought to protect. "To make our homes and civilization permanent is to guard our topsoil, keeping it fertile and productive," he concludes. In the closing paragraphs of the Author's Preface, Faulkner's apparent belittling of American farmers still rankles, as Jack takes one last shot at undoing presumptuous comparisons with farmers of other nations: "We are not Pagans of the soil, but tillers of it, and as such, we must take care of it and it in turn will take care of us."

Jack's position vis à vis Faulkner was unequivocally that the author of *Plowman's Folly* had been co-opted by business agriculture and government-funded, land grant universities. Faulkner, a one-time Smith-Hughes teacher funded by the feds in their desire to reform and urbanize country life through cultural re-education of the nation's rural residents, embraced the scientific paradigm then endemic to all levels of government. Then as now, it seemed to take an outsider to both these cultures, Academe and Corporate Farming, to critique both, to advocate for the average midwestern tiller who operated by conscience and common sense rather than by the latest county extension bulletin or academic paper. "Now and then," Jack writes in the opening chapter of *The Furrow and Us*, "here and there comes an aggressor to the vital natural laws, a portent to the natural order, proving his point with smooth language and, before a sobering thought has time to germinate, many accept the new idea against their common sense." The germination metaphor the "dirt-farmer" chooses here to oppose Faulkner and his allegedly insidious interests, is noteworthy, as it suggests that Faulkner fights on the wrong side—the side Joseph Campbell and George Lucas would immortalize, some dozen years after Jack's death, as the *Star Wars* Dark Side. Like Lucas and, indeed, the southern agrarians who collaboratively published *I'll Take My Stand* in 1930, Jack aims to unmask the treachery of institutional man and his betrayal of the independent farmer at the expense of corporate interests. This same cause, along with its missionary flavor, would be taken up decades after Jack's death by farmer-writers like Wendell Berry and Gene Logsdon, who, in their respective volumes, *The Unsettling of America*, and *At Nature's Pace*, would decry the land grant university's betrayal of the small-time producer. Jack's words are worth quoting here at length:

> Small wonder, then, that titled educators take a special interest in this branch of farm business, for farm economics is the brainchild of our national planners, and their prognostications can be dispensed from the round table with dignity.
>
> Not so with the problems of the soil. To master its meaning is to live on it, live by it, ever watchful for any deviation in response due to variations in tillage practices, getting the feel of it in the spring and working and observing it throughout the changing seasons. (60-61)[2]

One can almost hear—in the cathedral of agriculture where Jack and Faulkner each took a turn preaching their respective gospels—a quiet but resolute *amen* coming from the farmers crowded into the backbenches.

To appreciate fully the work of Walter Thomas Jack—so much a product of the Corn Belt and its golden age of optimism and expanding markets, an age that increasingly viewed food as both commodity and weapon—is to put him in a larger historical and cultural context. In the way that salad days, be they individual or national, seem always to entail fateful encounters (one cannot help but think of Bill Clinton as a child shaking hands with John F. Kennedy), it should not surprise that Walter Thomas Jack and his family knew what would become the most celebrated Quaker family in the United States, if not the world: the Hoovers. W. T. Jack and Herbert Hoover, called "Bertie" by Jack and his family, grew up just a few miles down the road from one another in West Branch, Iowa, a fact recalled by Jack in his short memoir, "Between Two Worlds."

Walter Jack grew up in a community of Quakers, as did Hoover, that established their own schools, conducted their own trade, and buried their own dead. The West Branch Quakers, many of whom had migrated from Pennsylvania and Ohio with the opening frontier, steeped their young in traditional Quaker values: education, pacifism, and devout service. The Jacks, in fact, lived on a small farm just a few miles north of Hoover's birthplace in West Branch, a burg bordered on all sides by an ethnically diverse population of immigrants. When the 1900 Census-takers arrived at the Jacks' door, they found, in a small house mortgaged to John Jack, a family of five: John, who listed his profession as "day labourer," Walter and his sister Mae, both at school at the time, and John's wife Margaret and her father-in-law, William Forsythe, a former captain in the Confederate army who claimed to have served under Lee in the Army of Northern Virginia. Margaret Jack, who spoke still with a genteel, Southern lilt in 1900, grew up in an abolitionist household in Virginia that, according to Walter, offered permanent shelter and sanctuary to at least two slaves during the war. Still,

Margaret and her father, Virginians through and through, considered themselves southerners and devotees of the Lost Cause.

Walter Thomas Jack would grow up "Between Two Worlds," as the title of his short memoir put it: a Quaker son who distanced himself, in young adulthood, from the sect in favor of a more general Christianity, though he admired the Quaker's "Puritanical grace." Raised in a church that forbade singing, Jack could carry a tune and accompany himself on the mandolin; raised in a faith that prohibited dancing, Walter Jack was known for hosting the biggest and best Farm Bureau dances in Cedar County, where he would take up farming after marrying 26 year-old Amber Jane Pickert in a small church in Lisbon, Iowa in 1917. In marrying Pickert, who was of Welsh, English, and, primarily, German stock, Walt Jack did the unthinkable for a Quaker son: intermarried. In "Between Two Worlds," Jack, reflecting on the raucous, joyful funeral of Irishman and West Branch horseman Patrick Donovan, pens this fair distillation of his divided consciousness: "And, in my heart, I had the secret thought that there were two sides of life—one of gloom and one of fascination."

"Between Two Worlds" is punctuated with memories of death, ceremony, and religion, three topics that defined for young Walter the severity and the sublimity of Quakerism. He wrote secondhand accounts of Herbert Hoover's father's funeral. Herbert Hoover's father, Jesse, died in 1880 of heart trouble and other complications, leaving a $1,000 life insurance policy and a little property.[3] Hulda Hoover, who Jack calls "widow Hulda," in his brief account, supported the family by preaching and sewing. Jack recounts the community's suffering, at the time of Jesse Hoover's death, "a biting curiosity as to what would become of little Bertie (Herbert) the orphan, because, at six, he'd played, not worked, in his father's shop." Jack continues, "And the sentiment was unanimous among mourners, 'Too bad Jesse [Hoover] could not have lived long enough to have taught his son the trade.'" Hulda Hoover would die of pneumonia and typhoid fever just three years later when her son was nine years old, and nearby relatives would raise young Herbert until such time as he would leave for Newberg College—a small Quaker academy in Oregon run by his uncle—and, upon graduation, to Stanford.

Meanwhile, Walter Thomas Jack, who was listed in the 1900 Census as John Jack's "adopted son," would discover that he, like Hoover, had been an orphan—born to unknown parents and adopted by the Jacks from a Des Moines orphanage. Like Hoover, Jack would reach maturity in a surrogate household; like Hoover, Jack would fair well at academics, graduating high school with high marks and accepting a teaching post at a country school in West Branch, Iowa before courting and marrying Amber Pickert from nearby Lisbon. The Pickerts, fearless New York pioneers who had arrived in eastern Iowa well before the Civil War, were farming people; Jack's union with their Amber would mean for young Walter both a further distancing from his Quaker heritage and the acquisition of a lifetime profession: farming.

Both men, Jack and Hoover—separated by little more than fifteen years and united in calling the same Iowa hamlet home—would make their name via agriculture, directly and indirectly, respectively. Hoover first gained national attention for engineering food relief for the Belgians after the Germans invaded in 1914. Credited with saving thousands of Belgian lives, Hoover gained the approval of Woodrow Wilson, who appointed him head of the United States Food Administration, which had broad powers over the prices, production, and distribution of food. In short, Herbert Hoover increasingly controlled agriculture: Walter Thomas Jack's world.

Jack's environmentalist thinking evolved quickly after 1917, when he took up the plow as a new, untested farmer. While his early ideology made explicit the comparison of a farm to a "battleground between man and nature," the Depression made Jack more circumspect. "After serious thought," he writes, "I decided I had inadvertently expected too much of mother earth, and, instead of mining her resources, I would study her way of doing things, listen to her hunger signs, and appreciate the fact that she mothers us all." In so saying, Jack—conservationist, pacifist, and agriculturist—braids three crucial strands of his maturing ideology: the same strands that Hoover, despite his Republicanism, represents. In describing his revelations in his Author's Preface to *The Furrow and Us*, Jack recalls being told by an old Quaker as a boy, "If you try to beat nature, nature will beat you." Hoover shared the same soil, and doubtless heard the same sentiments.

The second theme that develops in Jack's rhetoric and finds common ground in Hoover policies is a commitment to food as a commodity and agent of supremacy. While Hoover rode the coattails of his popular and well-publicized food relief programs to the presidency in 1929, Jack, a new farmer, was discovering age-old notions of agrarian superiority, biases made new by the American republic but gleaned, in turn, from the Romans—Virgil and Lucretius—whom Jack quotes with an ease surprising for a self-professed Iowa dirt farmer. Echoing the words inscribed by fellow Iowan, Regionalist, and agrarian Grant Wood on a mural at that time being painted in the library at Iowa State University (then the Iowa Agricultural College), Jack declares, "Where farming starts, other arts follow" and later, in a titular reference to man's status vis à vis nature: "First the furrow, then us."

On the heels of two world wars that pitted America's industrial and agricultural prowess against the rest of the world's, both Jack and Hoover begin to relate such a previously abstract notion as caloric production to

American preeminence. Hoover's meatless and wheatless days, amounting to a program of self-rationing to help the war effort, made "Hooverizing" synonymous with victory. Similarly, in the first chapter of his 1946 treatise, Jack takes up the cause of proteins—the caloric fuel behind the war's human engines that needed to be produced at "maximum climatic levels." It became Jack's goal to discover, with a Quaker's modesty and forbearance, what those maximums were—how much mother nature could produce without going bankrupt. Not until the 1960s, with the publication of such classics as Ralph Ehrlich's *The Population Bomb* and Francis Moore Lappe's *Diet for a Small Planet*, would the issue of sustainability be so poignantly written. Risking book sales and the further scrutiny of his tradition-minded neighbors, Jack challenges the farmer/reader of his book with a series of rhetorical questions about their own stewardship or lack thereof, concluding with a cautionary note: "Maybe you have farmed out this fertility faster than you have replaced it, and if you have, don't charge up declining yields to luck or old mother earth."

In the mid 1940s, when Walter Jack's pleas were widely read, but largely unheeded, American agriculture enjoyed a post-war boom spurred on by new chemicals developed during the war, such as the insecticide, DDT, later determined, according to historian David D. Danbom, to attack the nervous system, and 2,4-D, a herbicide based on a growth hormone. Soon added to this toxic array was Anhydrous Ammonia, a new nitrogen fertilizer that could be injected into the soil, an application that environmentalists and ecologists warned would lead to nitrogen contamination of groundwater and, more generally, to poor, band-aid farming methods.

While the operations of many of his fellow farmers grew precipitously, Walter Thomas Jack sounded the alarm, not by sticking his head in the sand, but by, as the saying goes, doing as the Romans do: in this case, applying heavy nitrogen despite his reservations. In the mid-1940s, chemical-based farming of wheat, rice, and many other grains increased production by a heady 50 percent, according to Danbom. In his thoroughly readable history of rural America, *Born in the Country*, Danbom reports that net farm income rose as much as 156 percent between 1939 and 1944. Against this backdrop, Walter Thomas Jack, ever the amateur natural scientist, decided he would put the nitrogen theory to a farmer's practical test. To the shock of fellow neighbors like Mechanicsville's Forrest Johnson, who remembers going out to the Jack farm to see what kind of madness was afoot, Jack selected a perfectly good field of black Iowa dirt and commenced with the unthinkable. First, he conceded the use of fertilizers to please his detractors, though he hoped to show that fertilizer itself could not make up for overburdening the land. Purchasing 400 pounds of 2-12-6 commercial fertilizer, he proceeded, in the spring of 1944, to divide his otherwise good field into three test plots:

one with a normal stand of corn plants per acre, another with nearly double the normal stand, and still another with nearly triple the usual number of corn plants.

Though the financial cost must have been tremendous, Jack proved his theory, namely that not even nitrogen could mask the damaging effects of overcapitalizing the soil. Suffering yields as low as 70 bushels per acre in the field planted at nearly three times carrying capacity, Walter Jack sacrificed his pocketbook for confirmation of a natural law—one that his nitrogen-drunk neighbors could now see for themselves. Encouraged by his definitive experiment and its appropriately sobering effect, Jack begins *The Furrow and Us* on an emphatic note: "Ever since the earth was created the laws of nature have been misinterpreted, revised, distorted, and manipulated to please man's vanity, his religion, and his aversions. . . . In one short stroke he vainly tries to change her habits that she has successfully practiced since the beginning."

In undertaking this field experiment in willful gluttony, W. T. Jack did not condemn or parody his fellow farmers. Instead, he made tangible his own early failures. Remembering the dark days of the Depression, when grain surpluses plagued Washington and the only way out for a capital-starved farmer was to grow more grain on existing acres, Jack admits that he had been "too busy worrying about bad luck and low prices to take stock of myself." Further detailing his ignorance in hindsight, he writes, "I had done the job of plowing, planting, and harvesting in the usual manner, unmindful of the retribution that follows when there is not the proper relationship between the tiller and the tilled. *The proper relationship between the tiller and tilled* encapsulates, as a phrase, Jack's world view, explaining why, even though he bucked the trend towards chemical-intensive farming embraced by many of his neighbors, he ultimately stood beside them. True to his Quaker roots, Jack resolved to educate himself out of worry and to sensitize himself to the laws of mother nature, who, he understood, would ultimately take care of him if he took care of her.

In standing by his practical brand of field science, Walter Thomas Jack stood for the lay farmer, who, he believed, was a true scientist. Reminding his reader that the greatest scientists did not always regard themselves as such, he characterizes discovery as a creative, serendipitous, even spontaneous act—"quite by accident, formulated laws . . . destined to become basic scientific facts." Citing scientific improvisers Marie Curie and James Watts, Jack illustrates his point with an appropriately agrarian metaphor, declaring, "a laboratory can be an idea, dream, plot of ground, or, in fact, the layman's work." This natural laboratory of the practical imaginer and farm-grounded fabulist Jack contrasts to the university's "pretentious grills" and "spotless rooms where technicians manipulate mysterious devices."

Research, argues Jack, ought not to be confined to the classroom or laboratory but should instead be practiced by "all classes of workers." Here, once again, Jack counters his old, no-till nemesis, Edward Faulkner, the Smith-Hughes teacher of agriculture and a self-made, so-called soil scientist. Not only was Faulkner's theory erroneous but so was Faulkner, Jack suggests, because he was not a man of the soil. Faulkner, tone-deaf to the soil's rhythms as Jack heard them, would, Jack thought, suffer nature's revenge for his estrangement. "She [nature] will call these perpetrators to account with the imposing of the usual penalty, declining yields," Jack intimates. But W. T. Jack's most serious dig against Faulkner—that he was pawn, as he saw it, of academic and commercial agriculture—comes midway through *The Furrow and Us*, where Jack intimates how he feels after reading *Plowman's Folly*:

> So convincing was this tale of woe that I began to look upon myself as the progeny of a mountain goat or a glorified ape. Then, quite suddenly I became aware that these animals would not, or could not, farm for thirty years, raise a decent family, and contribute something, even though a mere trifle, to civilization and culture. No, in this case, if there is a goat involved, the scent emanates from the direction of the writer of such stuff. (50)

Echoing Thomas Jefferson's foundational belief that farmers are the most virtuous citizens, Jack updates and makes more ambitious the notion of agrarian superiority: "It is certain that soil fertility and health are conjunct factors in the scheme of life, and the farmers of the future will be even more important to national health than medical men."

The effect of the Jack vs. Faulkner controversy concerning tillage and no-tillage, what *Time* magazine called the "hottest farming argument since the tractor first challenged the plow," would echo for more than half a century. While Faulkner's work would propel him to national and international fame, prompt several additional books, and secure him a financially sound future as a soil consultant, Jack's book would go underground—its limited print run passed hand-to-hand between family farmers, college professors, and eager agriculture students for more than a half century. In the definitive, bibliographic series edited by Wallace C. Olsen, *The Literature of the Agricultural Sciences*, Oregon State University's B.P. Warkentin includes special mention of what he calls "subjective" portrayals of the soil like Faulkner's and Jack's, citing Von Humboldt's quotation by way of context: "In order to comprehend nature in all its sublimity, it would be necessary to present it under a two-fold aspect, first objectively as an actual phenomenon, and next subjectively as it is reflected in the feelings of mankind." Before he leaves the subject, Warkentin notes that books like Faulkner's and Jack's, reflecting ideas "so common to our heritage, of the unnaturalness of urban life, and the purity of the rural life" enjoy widespread, enduring popularity that continues to underlie "the concept of the independent yeoman farmer and the efforts to preserve the farm" as well as, more generally, the increased currency of literature of the environment and of place.

In the final analysis, neither Faulkner nor Jack proved entirely correct. Faulkner's method of no-till would take a long time to catch on and would never achieve a clear majority. Jack, aiming to make a popular splash to rival Faulkner's, would have to content himself with a mostly specialized, regional readership and an enduring family legacy. Jack's only son, Edward Lee Jack, would go on to win national prizes in cornpicking and combining, along with a half dozen Master Corngrowers awards for high yields, all the while continuing his father's obsession with the perfect ear of corn produced in well-nourished soils. The extravagant, experimental test plot that so shocked W. T. Jack's neighbors in the spring of 1944, would become a mainstay for Edward, who would routinely plant, up until his semi-retirement, as many as seven hybrids in order to determine the one best suited to local conditions. Michael Jack, Walter's grandson, would, somewhat ironically, become one of the first in his area to try Faulkner's no-till method in large scale, winning a Master Corngrowers award for high yield and ASCS conservation honors in the process. But, in accepting the award from the Corngrowers, Michael Jack would grant that such high yields required healthy doses of chemical nitrogen—the very component his grandfather Walter resisted.

In the culture at large, Jack would trailblaze for subsequent generations of environmentalists, who, using Jack's logic, would resist genetically modified corn and soybeans along with growth hormone-raised livestock. Jack's chapter, presciently titled "Hidden Hungers," anticipates, in a mere half dozen pages, many of the major agricultural and bio-cultural trends of the last half of the twentieth century, including increased vegetarianism ("Mother [nature] transmutes the dross of earth into energy that can only reach us through the green leaf.), natural pharmaceuticals ("In this . . . earth are the essential chemicals that go to make up the formula that has baffled scientific men . . . and is . . . the key of life.), and dependency on vitamins ("*We are what we eat* . . . is a quotation that is . . . sending many people to the corner drugstore for their daily healthy doses."). In defining the destructive cycle of so-called hidden hungers, where "animals suffer . . . and humans, in their turn, consume the products of starved plants and animals," Jack looks forward to future consumer demands for free-range and grain-feed animals. Eerily, Jack even anticipates the growing incident of infertility among men and women in industrialized countries, citing studies linking calcium-deficient soils to impotence in both cattle and rabbits. "Nature's warnings allow time," Jack simultaneously comforts and cautions, "for us to heed them against our own extinction if we will look to the soil and conserve its fertility against the forces of exhaustion."

Of course, Jack proved wrong about some things, making apropos Auden's elegiac words for the great Irish poet William Butler Yeats, himself a product of the country. "You were silly like us," Auden eulogizes in his poem "In Memory of W.B. Yeats," "your gift survived it all." In print, for example, Jack underestimates the impact of Darwin, believing that the great scientist neglected diet as a transformative evolutionary agent. In places, and consistent with his era, Jack's writing puts undue emphasis on genetics and breeding, as it posits, among other whimsical hypotheses, that midwestern farmers' sons are more fit for military service because of superior diets. Elsewhere, Jack falls victim to outmoded notions of superiority of the first over the third world, and of man over beast.

However, what should impress the twenty-first century reader of *The Furrow and Us* is just how right, how dead right, Jack's prophecies proved, and how courageous was his stand against big agriculture. Like the best speculative and science fiction, Jack's chapters offer the careful reader dramatic prognostications—in this case, a future utterly transformed by diet. In form, Jack's essays, part memoir, part science, and part "invaluable reference," as the dust jacket reminds, serve as a harbinger of the renewed popularity of nonfiction in general, and of memoir in particular. As a farmer writing with an unabashed farmer's sympathies and sentimentalities, Jack resolved to make his story a real one, replete with the natural dramas inherent to a life lived close to the soil.

Most important of all, though, is Walter Thomas Jack's enduring and edifying spirit, a quintessentially Transcendentalist and, therefore, American union of the pragmatist with the poet. Walter T. Jack loved farming, and in this love are sown the twin seeds of science and spiritualism. Jack's embrace of his identity as a lay farmer-writer puts him squarely in the camp that includes Thoreau—whose recording of his intimate ties with nature and agriculture made his literary reputation—and Robert Frost, likewise an autodidact, who, like Jack, made his living as a farmer and a schoolteacher. Closer to Jack's home in the Midwest, Carl Sandburg offers a compelling analog. Though identified with the city of Chicago, Sandburg grew up in Galesburg, Illinois, a small, agrarian outpost less than one hundred miles from Jack's West Branch. Sandburg, like Jack, wrote workaday nonfiction for newspapers and magazines as well as fanciful children's tales like the

Rootabaga Stories (1922), a sequence not dissimilar to Jack's unpublished volume of farm-based fables, *The Old Farm: Stories for All Children.*

Compared to Sandburg's, Walter T. Jack's modest fame seemed almost accidental, avocational. True to his Quaker heritage, Jack cultivated not only his corn, but a paradoxical modesty that allowed him to claim proudly the title "dirt-farmer" and to include self-effacing comments like "Thanks, brother, I am only a farmer" in deferential response to his book's sophisticated, sometimes acrobatic lines of rhetorical questioning. Yet his conclusions were, in their modesty, revolutionary in their observation of natural law. Edward Faulkner, by way of contrast, owed his fame to an ambitious, entrepreneurial spirit and a set of novel ideas about no-till borrowed from agriculture universities. The growth of Jack's popularity over the last sixty years—steady and organic—closely parallels the natural growth he encouraged as a yeoman farmer. Beyond the uncanny predictions of vitamin supplements, third world hunger, organic farming, soil depletion, and human and animal infertility, Walter Jack will be remembered as a writer, who, even in his accidental fame, knew—profoundly knew—his place. We hear his certainty, his exquisite thanksgiving, expressed in the passage below, words that, with this new edition, happily reach new generations now finding a place on the land:

> For every person there is a spot, a place where they pledge themselves to return some day, because they were happy and contented there. . . . My place is under Iowa skies, floored with the world's best soil and walled in by the emerald skyline of native oaks. I have planted and harvested these fertile fields for many years and have learned from old mother mature that she holds no grudge. She has healed the wounds and scars I thoughtlessly made on her contour and patiently replaced the fertility I so lavishly spent. No, this is not agricultural science, but it does have to do with the story of the soil I have set out to tell.
>
> This sentiment does not draw its substance from my imagination; I have lived it—it is real and everlasting. (73)

PART ONE

Sentiment

Chapter I.

THE QUAKER'S NATURE: A PREFACE[1]

When I was a small boy, an old Quaker once told me, *If you try to beat nature, nature will beat you*. This bit of philosophy has been axiomatic in my agronomical thinking ever sense.

Man is the only natural enemy of nature—he fights the soil and the soil in turn fights back with its only weapon, unproductivity. When all people of the earth are convinced that all soils in every land are governed by the same natural laws, then mother earth can smile again and freedom from want will be assured. It is an admirable gesture to strive for better implements and more advanced ways of doing things, but when our enthusiasm conflicts with basic natural laws, it is well to consider what the continued conflict between man and nature will bring about.

Into this soil-picture the farmer fits as an equal part, for surely the soil can never be better than the tiller. This fact is becoming increasingly evident as time is constantly extracting new secrets from mother nature.

Time was when a parent dedicated the brightest member of the family to a professional life but rested easy on the more backward child, for the farm was always open to them, where they could earn their living by sweat and brawn. Today the situation is reversed, for to be successful and to handle the various intricacies involved in soil management, crops, and livestock, the modern farmer must be a creative thinker and realize that their farm is but a scientific laboratory. To realize the most from it, the farmer must be keen to observe results from personal experiences and be ready to capitalize on the experiences of others. There is a singular appropriateness in the quotation of the old Quaker, and, while it is expressed in a homespun manner, its truth must form the framework for intelligent management of our soil.

The recent trend toward subsurface tillage is a transgression of one of nature's most vital laws. While comparative yields indicate no great reduction in one season, the soil, where robbed of natural processes, will register a steady decline in productivity, and if continued for a long period, the net result can only be soil depletion.

The most ardent critic of the furrow, Edward Faulkner, blandly ridicules the American farmer for not producing yields comparable to the Egyptians who still use simple agricultural tools.[2] But those living along the Nile know nothing of the law of the minimum, because the river on its yearly rampages spreads the valley with a lavish layer of balanced soil nutrients.

No, the American farmer does not envy the farmers of any country for they know the furrow is mightier than the Nile—it has no boundaries and it is turning the wilderness into permanent homes for the ever-increasing population. Civilization has continued to follow the plow—it has conquered the wilderness and has worked hand-in-hand with nature in reestablishing marginal lands, thus spreading the boundaries of human habitation.

Perhaps we ought to be discussing not just the civilization-making effects of the field plow but also the garden trowel, for it, too, breaks the soil and stirs humus to action.[3] It is not the object in these pages to belittle the small experimental plots such as Faulkner's that spring up in profusion over the entire country. They have, indeed, given much interesting and valuable data governing plant behavior under various conditions of soil, tilth, moisture, and climatic conditions. Yet, the small plot represents but one type of soil and one elevation, and conclusions attributed to its performance may not be representative of a large field that runs through low-lying black loam to high clay soil. For definite opinions on methods of soil management, it stands to reason that a large field, embodying different soil types and elevation, will yield more comprehensive results for the farmer.

I have just such a field—80 rods wide and 160 rods long. The first 40 rods run through a fertile flat of black silt loam. The next 40 rods run to the top of a clay hill. The remaining 80 rods descend to a flat comparable to the other end. This clay hill, in extremely dry periods, usually resembles a scorched dome resting on a green base, as the crops dry out in hot weather, and, in spite of the extra care they receive, fail to produce a paying harvest.

During the years when horses supplied the farm power, the plow would invariably turn a deeper furrow in the moist bottomland than the high ground. The high clay soil plowed much harder, and, in pity for the horses, the plow was raised a notch. This shallow plowing created a plow-sole—a hard, impervious soil structure that permitted little moisture to penetrate. I realized that this hard furrow should be broken up, but every year the same power problem presented itself with the resultant poor crop yield.

The tractor solved my problem, and with its steady power, the old hill was soon plowed to the desired depth. I now look out over this field, where, after a hot and dry August, the corn is uniformly green with the prospect of an 80 bushel per acre yield. But his transformation did not just happen overnight, nor was the deep plowing the only factor responsible. This high ground had been tilled shallow for so long that it was almost devoid of humus, and, consequently, it was near depletion. Erosion exacted a heavy toll, as little water could penetrate the solid soil, thus leaving a greater portion to run off.

The other key to my field's transformation was an ideal cover crop, a cover crop being equally essential on sleeping farmland in the winter as on the household bed. This field was in red clover in 1944, and I let the lush growth remain on the ground. This thick mat of clover proved not only an ideal cover crop that preserved the topsoil, but it proved a valuable source of organic matter to plow under the next spring. If the plowman has a serious folly, then, it is the time he plows, *not* the act of plowing.

No one has paid proper tribute to our topsoil.[4] No one seems to recognize the fact that the furrow envelopes this priceless source of life. From the few inches of topsoil, all plant and animal life must generate, and once generated, must supply the vital forces to perpetuate their kind.

Little wonder, then, that the creator of the universe endowed nature with the economic distribution of the critical soil elements that allows but a small part to be released each year. To force nature to yield up more of these elements in a given time would turn the natural balance of food and feeder into an unbalanced state of survival of the fittest.

To make our homes and civilization permanent is to guard our topsoil, keeping it fertile and productive, for if we fail to do this, we disintegrate.

And so it is—first the furrow—then us. When we turn the soil and accept its bounties, we are not robbing the land. We need not fear retribution. We are not infidels of the soil, but tillers of it, and as such, we must take care of it, and it, in turn, will care for us.

This work is not intended as a paradox.[4] Its purpose is to prove that a permanent agriculture must be built around proven practices that are inspired by the natural behavior of soils and plants. *The Furrow and Us* passes on to its readers the results of many years of careful observations of natural laws and the responsibility we farmers have in guarding soil fertility so that it will continue to be a source of life and pleasure for all generations to come.

Chapter II.

THE FORCE OF CIRCUMSTANCE

Back in 1917 I embarked on an agricultural career as a life's work and, on a fertile Iowa farm, soon found that the farm was the battleground between man and nature.[1] Thus, the facts and conclusions set forth in this work did not come easily but were born of necessity, the product of the tedious trial and error method practiced by the tillers of the soil.

The going was good during the lush war years, and the few succeeding years put no tax on personal ingenuity, for prices were good, and if yields faltered and fluctuated somewhat, the gap between actual yields and maximum yields was spanned by high prices. There seemed to be no need for worry; old methods of doing things seemed quite adequate, and the complications of the chemical nature of the soil seemed to be of no immediate concern. The soil had always responded to our call.

When the war clouds of World War I finally cleared, I, like many other young farmers, was not ready to meet the challenge of diminished soil and falling prices. Soon the black clouds of economic disparity loomed up on every horizon, stealthily boring in, gaining momentum, until hard times peeled the shirts from our backs and swallowed up our savings. Many farmers who had purchased a farm by using up their lush-time savings as a down payment were forced to relinquish their safety net. Some of us held on through pride and an unwillingness to accept defeat, and we did the only thing we could to meet a fixed expense—increase production. This, of course, was the wrong thing to do, as agricultural surpluses were already plaguing Washington.

Our head-above-water mentality had its own repercussions, and with a greater percent of farmland under cultivation, it became evident that the soil was being burned out. Organic matter was being consumed by the crops more rapidly than it was being replaced, resulting in a sharp decline in yields. Low prices and poor yields, we knew, would spell ruin for us all.

I reasoned, as did many growers, that while we could not raise prices with surpluses, we could cut down operating expenses by producing our requirements on fewer acres. After serious thought I decided I had inadvertently expected too much of mother earth, and instead of mining her resources, I would study her way of doing things, listen to her hunger signs, and appreciate the fact that she mothers us all. We tillers enjoy the feeling of pride and thanksgiving that has blessed our efforts as we look over a lush meadow, billowing fields of grain, and verdant pasture. This is a heritage, not to everyone, but to those who feel themselves subordinate to mother nature and are willing to follow her teachings rather than direct her thoughts.

With increasing interest in the natural phenomena of the soil and the growing plant, I saw that the crops began to reflect the value of my study and cooperation, and I observed the dawning of a new day new era in farming, an era full of hope and assurance that our soil can be kept well and made even more productive.

During those years when time and profits made it necessary to do things as economically as possible, I used the disk instead of the plow to prepare the seedbed for corn. The results were the same as most slipshod methods, and, after a few years of this practice, the soil was definitely drier, and a few hot days in the summer revealed definite deterioration due to the hard, dry seedbed. Soil worked to a shallow depth, I discovered, cannot trap and hold enough water to support a heavy crop. Taking into consideration the loss of organic matter necessary in the process of nitrification, it seemed certain that abandoning the furrow for subsurface tillage was an act against common sense.

After years of watching the soil respond to the cooperation of the tiller, and, in fact, producing greater yields than were produced a decade ago, we can safely conclude that our basic ideas of soil management are sound.

Praise for commercial fertilizers, then, is intentionally absent from this work. The work deals with permanent productivity; and commercial fertilizers directly applied to crops do not fit in my definition of a permanent agriculture. As a direct aid to row crops, chemical fertilizers are comparable to vitamins in the human diet—vitamins that are foisted on the public by the lilting words of professional advertisers. True, commercial fertilizers are composed of natural elements that all soils must have to be productive, but feeding them to non-leguminous plants, like most of our grain crops, is a waste of money. They build up false hopes, since they do promote luxuriant foliage, but for some reason no one can explain, ears of corn do not benefit from direct applications. These elements must be transformed into nutrients by legumes and soil agencies, and if this fact is kept in mind, the soil will acquire more organic matter by virtue of the stimulating effect of the chemicals on the legumes.

Despite these hunches, I was, at the time, hardly ready to get scientific with my claims against subsurface tillage and needless fertilizers. I had been too busy worrying about bad luck and low prices to take stock of myself. I had not thought that inefficiency might have had something to do with my plight. No, I had done the job of plowing, planting and harvesting in the usual manner, unmindful of the retribution that follows when there is not the proper relation between the tiller and tilled.

I was not aware that my shallow tillage methods had created a hard, impervious soil that would not store up moisture. I had not inquired into the reason why my corn fired sooner in hot weather. I was not concerned with the gullies that were cut, even on gentle slopes, and instead of sodding them, I performed the usual predatory operation of plowing them in, increasing the valley in width and losing tons of precious topsoil. I had noticed that the soil on my farm stayed wetter in the spring; the water would run off the gentle slopes or stand in puddles on the level ground, not soaking in, but I did not have the courage to make further diagnosis.

No, I was content to enter these items on the left side of the ledger—and in red. And as might be suspected, the longer I was content, the redder became the left side of the ledger. I had looked upon myself as an ordinary farmer, and this esteem was not surprising, for I had always kept abreast of seasonal jobs, and many times, before the advent of the tractor, had prepared the ground, planted, and cultivated one hundred acres of corn without help, not to mention small grain and hay crops. Little wonder, then, that I had but little energy left for serious thought of the teeming universe deep in the earth that was ready and willing to work for me if given the chance, nor of the vital natural laws of which I knew little.

The force of circumstances was to change all this. I found that a few moments of the day spent in observing natural phenomena and a few moments in the evening reading the experiences of others became an interesting and profitable habit. It gave me a sense of security and freedom from worry, for, after all, worry is the result of inefficiency.

Readers are advised to walk through their own fields or plots and to reach down and grasp a handful of soil. Is it as dark in color as it used to be? Examine it closely and see if undecayed pieces of stalks or other healthful particles of organic material appear as desired. If not, maybe fertility has been farmed out of the soil faster than it has been replaced. If so, charge up the declining yields to the person in the mirror, not to poor luck or old mother nature.

Chapter III.

In the Beginning[1]

Ever since this earth was created the laws of nature have been misinterpreted, revised, distorted, and manipulated to please man's vanity, his religion, and aversions.[2] They have been altered to fit his own likes and dislikes, and, in one short stroke, he vainly tries to change nature's habits as successfully practiced since the beginning. To gear nature to our modern civilization is but courting madness, but to tune our needs and desires to the forces that have developed our earth after creation is to insure for ourselves a better place to live and a permanent civilization.

Now and then, here and there, comes an aggressor to the vital natural laws, a person proving a point with smooth language, and, before a sobering thought has time to germinate, many accept the new ideas against their own common sense. These statements have an appropriate significance at this time when we tillers of the earth are called upon to question the value of the so-called subsurface tillage method lately sold to us by so-called agricultural "experts."

As guardians of the soil, we sometimes wonder of what use are the multitudes of worms and insects with which we yearly contend. After all, were we not given dominion over these natural agents that carry out nature's laws? We would do well to remember that when the earth lay in formative state, the very things that plague us now as farmers were important, even vital, for the crucial task they were to perform. The native rocks had to be broken down to form basic soil structure. The scant vegetation had to be worked into the soil to encourage the important soil organisms. Oxygen and water had to penetrate the soil if it were to accumulate fertility. All this groundwork had to be accomplished by mother nature alone. And all this had to be done, and done well, to prepare a fertile soil that must precede the coming of human races.

The universe of underground organisms—from minute bacteria to the large earthworm and on up to the larger burrowing animals—each played a vital role in building our precious topsoil. In a crude way, worms, insects, and beetles burrowed deep into the earth, and, through the fissures they made, oxygen followed and aerated the soil. Through the same fissures, water was able to penetrate, giving the soil organisms another source of life. Many of these larger denizens of the earth fed on plants growing on the surface, carrying certain amounts of organic material into their underground homes, supplying the minute bacteria with food.

Thus nature charted a course. The processes were and are simple. They show us the way our soil was built, how fertility was fixed into it, and how its productive capacity can be maintained. The natural processes were slow, with but slight increases recorded each year. But after many centuries, the earth took on new form, and, as fertility increased, new forms of plant and animal life came into being and, in turn, performed their function, each contributing something to make their presence useful.

In earlier times, certain plants would grow in the regions where soil conditions were to their liking, but today the soil can be balanced to meet their particular needs, and the benefits of foreign plants can be enjoyed by peoples from far-flung corners of the earth. Many of the original plants, after growing for thousands of years in their natural haunts, gradually used up certain soil elements peculiar to their individual needs and declined in vigor. While these species may have become extinct in their original environment, they gradually spread to soil that was virgin to them.

The replacement of worn elements was the greatest hazard with which nature had to contend, and this fact seems to be the logical reason for the changes in regional species of plants and animals as recorded in the prehistoric fossil record. That soil moisture was also of prime importance in base-exchange and in affecting soil solutions was amply provided for in the earthworms and various species of bugs, beetles, and larger rodents. Water, the universal solvent, followed the runways of these burrowing agents, soaking deep into the ground, making it moist and mellow for the effective performance of soil organisms. In effect, this served as a substitute for the mellow, absorbing earth made possible by the furrow, and the digging and boring of the worms and animals served as a primitive plowshare.

Why then is it necessary to go to all the work to plow the soil if the task is done by nature? Today, the demands made of the soil are much greater. We expect large yields of protein foods that are but a combination of soil elements. We expect to keep yields up to the maximum climatic levels, and the processes of nature are too slow to replace the great amount of soil nutrients that go into large yields. This replacement cannot be accomplished by chemical fertilizers if we expect a permanent soil fertility, for old mother nature only compromises on her own terms; she will not trade her natural soil air nitrogen processed by legumes for nitrates or sulfates without imposing her usual penalty—decreased yields.

Since Dutch scientist Anton Leeuwenhoek discovered the microbe, soil organisms have generated much interest in agriculture and beyond. We cannot live without them, for they generate life itself. Neither could we die a natural death without them, for they are directly responsible for many illnesses. But to the tillers of the soil, the chief concern centers on those friendly groups of soil bacteria that eagerly consume all organic material, turning it into free nitrogen that goes to build a fertile soil and the residue into available plant nutrients.

We are now justifiably concerned with the culturing of these important organisms in the soil, especially with turning the furrow down to their feeding zone. They are living organisms and must have moisture, air, and food. They cannot function in dry soil; therefore they cannot get at the residue material that is on the surface without our help.

Incidentally, the very fact that soil-building bacteria cannot utilize chemical fertilizers should be satisfactory proof that chemicals have no place in building a permanent soil fertility. Their value chiefly concerns the soil's need for phosphorus and potash. As a direct stimulant to crop plants, they should be avoided. This applies particularly to nitrogenous fertilizers. Soil organisms, like plants, have their soil preference, and whenever these organisms thrive and multiply in the soil, we need have no fear of our soil's productivity.

Chapter IV.

A Science Is Born[1]

If one class of workers could be eliminated without serious economic distress, which would you suggest?[2] If you suggest the farmer, you will be in error, for he is the indispensable worker and supporter of all other classes. We could, if necessary, get along without many classes of business people, politicians, lawyers, teachers, soldiers and even doctors—we did get along without them for many years. But suppose the farmers would all vanish quite suddenly—what then? The people of the world would starve before they could learn to raise their own food. Where farming starts, other arts follow. Even the first city could not arise were it not for the advances of farmers.

"Cultured" nations—a term suggesting a leisure class of educators, scientists, and writers—could not arise from an unproductive land or a land without farmers. Early records reveal humans living as wanderers, until their growing need for more food forced them to domesticate herds of sheep and cattle. Afterwards, these early peoples became herders instead of hunters, producers instead of predators. This change lessened their subsistence range and became the vanguard of a permanent agriculture. Finally, it was discovered that food could be grown from the soil instead of depending solely on herds and flocks.

Early peoples found that lowlands offered an easy and productive soil to work. There they could raise their food with a minimum of labor. Eventually, tribal warfare drove many clans to the hills, where they could farm less fertile soil in peace. In these highlands it was proven that even thin soil would produce crops if handled wisely. In these rugged climes, it was soon learned that some means must be employed to hold the soil in place. Hence, they became the cradle of terracing, strip cropping, and contouring. Many of the fertile valleys were exploited and depleted, thus only the records of a vanished civilization remain as a warning to future tillers of the soil.

The early history of agriculture might have but little significance to many, but, in reality, it is valuable chronicle of agricultural trial and error. "Primitive" people may have lived by myth and magic, but they observed natural events—noticing how, as if by some queer quirk, previously abandoned land became productive again. Early knowledge of the soil brought strength and importance to tribes whose tillers became more skillful with soil and domestic implements. And from these early and tentative histories, we draw worthwhile conclusions and formulate laws that make our progress steady and our methods safe.

Archeological research indicates that the earliest records of agricultural crops began during the Neolithic period, about ten thousand years ago. Human culture and advancement in agriculture seemed to run hand-in-hand, since the best-fed people attained a higher standard of civilization. Still, for thousands of years, progress in agriculture was slow, as tribal warfare occupied the time and talents of these early people, and instruments of war were of greater importance than implements of peace.

History shows that farming was at its best during the height of the Roman Empire some two thousand years ago. We get an important hint from Virgil as to the value of the crop rotation and the worth of legumes. He wrote:

> Where vetches, pulse, and tares have stood,
> And stalks of lupins grew, a stubborn wood
> The ensuing season, in return may bear
> The bearded product of the golden year.

Each agrarian era was highlighted by the work of a famous philosopher who indicated in his writing the advancement being made in agriculture, disproving theories of his predecessor. And so natural laws were observed and chronicled, many of which were proven untrue by later observers.

The coordination of biology and agronomy marked the real beginning of agricultural science and a realization that soil organisms must be cultured in the soil to insure dependable fertility. It is interesting to note the lamentations of Lucretius (900 BC), who complained that the world was growing old:

> She brought forth herbs, which now the feeble soil
> Can scarce afford to all our pain and toil.
> We labor, sweat, and yet by all this strife
> Can scarce get corn and wine enough for life.
> Our men and oxen groan, and never cease,
> So fast our labors grow, our fruits decrease.
> Nay, oft the farmers with a sigh complain
> That they have labored all the year in vain.

Time depleted the soil, but in turn built it up again, and the cause of this change was studied until the discovery of the relationship between soil organisms and soil fertility was clearly defined. The progress from this point was relatively rapid, and, with the broadening of the scope of knowledge of agronomy, agriculture slowly developed into the chief business of commerce.

Thus a science was born, and of all the words in the English language, none is respected so profoundly or misinterpreted so broadly as *science*. We are all scientists—accomplishing important work by certain knowledge. In light of this realization, we must stop regarding scientific work as the result of magic made by the magician-scientist of his supernatural power. In reality, science is only our ability to analyze problems in the light of natural law.

Many people would keep the layman in the dark as to the actual meaning of the words *laboratory* and *research*, as the layman has long held these institutions in high esteem. The layman's deference is not surprising, for the modern laboratory is a well-equipped institution presided over by specialized experts. However, the greatest scientists of all time did not regard themselves as scientists, but in following certain lines of thought, they sometimes—quite by accident—formulated laws that were destined to become scientific facts.

So a laboratory can be an idea, dream, plot of ground, or, in fact, the layman's work. It is not necessarily enclosed in spotless rooms where technicians manipulate mysterious devices. To Leeuwenhoek, it was a musty basement where he peered into the minute world of the bacteria. For James Watt, it was a kitchen and a boiling tea kettle. For Pierre and Marie Curie, it was a leaky shed with a dirt floor. Research is not necessarily confined to the classroom or specialized laboratory, but can be, indeed should be, practiced by all classes of workers.

Chapter V.

WHY WE PLOW[1]

The recent controversy demeriting the moldboard plow has led to many arguments and much speculation on the part of farmers everywhere.

In my judgment, the method of stirring the soil without turning under the top with its crop residues, or, in modern terms, subsurface tillage, was practiced by primitive people of every land since the beginning of time. The principle was outmoded with the advent of the woodboard type that turned only a portion of the surface under, since the wood surface could not be induced to scour. Only about one hundred and fifty years ago was the steel moldboard design put in general use. To it and its principle, we owe much.

The crude implement used by the ancients was, in effect, an object pulled through the ground to loosen the soil and make it tillable. There was no device on these primitive implements to turn the surface of the soil, with its growth of natural organic matter, or to stir the manures that may have been scattered. The implement left this valuable source of ultimate fertility on the surface, thus transgressing one of nature's most vital laws. This truly primitive method led to a rapid depletion of the soil. Only after it was discovered that soil-building agencies were living, soil organisms supplying fertility and tilth to the soil, was the present moldboard plow designed.

Civilization has ever continued to follow the plow; it has conquered the wilderness and has worked hand-in-hand with nature in reestablishing marginal lands, thus widening the boundaries of human habitation. Since the advent of our present method of tilling, the soil has shown a steady increase in fertility, and with the present knowledge of biological research, the old plow still holds first place on the agricultural hit parade.

Those hostile to present tillage practice point out that, since most of the nitrogen requirement of the plant comes from the air, there is no need to encourage the soil bacteria to supply the major portion, and therefore the organic matter in the form of trash and manure had just as well remain on the surface as a guard against erosion. This argument seems to be a radical departure from the true principles of agronomy.

To prove the worth of organic matter, or humus, in the soil it is necessary to know what goes on in the soil and the plant, and to determine the relation between the elements, the soil, and the air. While the air is approximately three-fourths nitrogen, and nitrogen is the chief constituent of protein, we must not deceive ourselves by thinking the leaves of the plant can absorb their requirements. The root of the plant is equipped with the mechanism necessary to gather the nitrogen from the soil, while the function of the leaf is to manufacture these crude elements into the proteins, carbohydrates, and fats that form the basis of human and animal food.

Therefore, the soil must be aerated to allow the free circulation of air through the topsoil, because, without oxygen, soil organisms cannot live and function. By plowing the soil, we incorporate the organic matter throughout the soil structure, where it keeps the earthy particles from forming a hard mass and permits aeration. This combination of conditions is favorable for the culture of soil bacteria and, consequently, for fertility.

Soil organisms are dependent on oxygen for the important role they perform in the prevention of carbon dioxide toxicity. Nitrogen fixation in the soil could not function without a combination of bacteria and organic matter. The root cells take in the soil nitrogen, and the residue, in the form of carbon dioxide gas, is exuded by the leaves. Therefore, the whole physiological scheme of plant hinges on the aeration of the soil, and the tiller of the soil must recognize this basic fact if he is to get full benefit.

It is definitely known that the chief soil-building elements added to the soil, in recompense for wear, are in an insoluble form and must be placed in the soil with organic matter that is in a state of decomposition. Decomposition forms an acid that, in the case of phosphate, liberates the phosphorus. An ordinary, prudent observer can detect the water-holding ability of a soil well-supplied with humus, and, accordingly, can note the advantages of these sponge-like pieces of material in soaking up water in wet times, releasing it to the plant when needed.

Every branch of industry must have imaginers if that industry is to make progress. The moldboard plow was imagined from the ranks of crude agricultural implements and today is the chief factor in keeping our food supplies plentiful. Those who, despite common sense, still demand empirical proof of the value of the old moldboard plow and the furrow, might well consider the results of the Corn Belt experiments.

The controversy started over the new theory spurred the Iowa Agricultural College to greater experimentation.[2] The results so far indicate that the only gain accorded to the new method was in labor saved, but this slight gain in time and labor was decidedly at the expense of yield.

Growing corn without the moldboard plow showed a power and labor saving of 20 percent, but the corn yield went down 11 percent. This experiment was conducted in five representative Iowa counties, and the net result changed little even after different soil types were accounted for. In Story County, Iowa, the experiment of the difference in costs has run for four years, but the yields were computed in 1944 and were the average of the yields in five counties, which should carry considerable weight. Disking the surface planting reduced work hours 17 percent and power requirements about 30 percent. Corn yields on land planted after the disk averaged 56 bushels per acre compared to 63 bushels per acre on plowed land. Hard ground listing, which is new to most sections of the Corn Belt, reduced labor costs and also power costs and did show a slight increase in yield.

As was suspected, fertilizer showed some gain in yield on the listed ground while it did not show any increase in yield on the plowed ground. This fact seems to bear out the conclusions previously stated—that the productivity of the soil is dependent on the organic matter that is turned under and processed by soil bacteria into plant nutrients. On average, when applied to all plots plowed and planted in the old way, the addition of fertilizer made no difference in yield. Both fertilized and unfertilized plots yielded an average of 63 bushels per acre.

On subsurface-tilled land, fertilizer increased the yield from 55 bushels to 58 bushels per acre. This slight increase would yield no advantage above costs and the extra labor of handling. It must always be assumed that where nitrogenous fertilizer is used there is an acid reaction in the soil, and 100 pounds of commercial nitrogen will, so to speak, rub out 500 pounds of lime, leaving a toxic condition that is harmful to growing plants. Indeed, much of the soil in the Corn Belt is not suited to the listing method of tillage and will not respond to any type tillage implement other than the moldboard plow. Legume and bean fields and stalk ground showed greater possibilities of yielding fair returns when listed, but a continued disregard of the value of plowing under all crop residues, especially on the heavy Corn Belt soils, will show steady soil depletion.

Already, some manufacturers have on display the subsurface tillage machine, but time, and time alone, will attest to the success or failure of the system. If they prove successful, and if the soil can be kept aerated and fertile, then all that I have learned during the past thirty years will have been overturned, and nature's most vital law will have to be repealed.

The result of any worthwhile experiment dealing with radical changes in doing things, especially where soils and plants are involved, should not be judged by a single year's experiments. This new tillage method will continue for many years, not only by experiment stations, but by farmers, who, faced by labor shortages, will try the method out thoroughly even at the risk of lower yields.

The corn borer is another factor to be considered seriously. Since this pest is a stern reality, farmers are urged to turn under all stalks and surface trash to a good depth, since borers cannot survive without air. If the borers are left to their own devices, it means another hazard added to the already overburdening list with which the farmer must contend.

It is pointed out by sponsors of subsurface tillage such as Edward Faulkner that nature, if left to her own devices, keeps the soil fertile by leaving all residues on the ground. No-tillers are quick to point out that soil not disturbed by the plow holds its natural physical proportions better. This reasoning is purely academic and has no logical proof. Domestic plants, for instance, cannot compete with the wild plants of nature. They are the children of the wild soil, and as such, make different demands on the soil. The domestic plants are called upon to produce abundantly, and they cannot perform this function without the farmer's aid and intervention in the survival of the fittest. Therefore, it is logical that to meet their needs they must have a deep, aerated seedbed where the laws governing soil and plant can work freely. The natural virgin cover must necessarily be broken to make way for our food plants, but the cycle of soil-building agencies must also be recognized, and the plowed-up state of the soil must not exceed the limit of soil organisms and humus.

We know now that the processes of nature are too slow to feed a hungry world. It is the province of the tiller to speed up the process of soil building, incorporating into the earth the minerals and material that have been withdrawn. Tillage methods cannot in themselves build fertility in the soil, but they do set the stage for the successful performance of the natural soil-building agencies. The larger the stage, the more satisfactory the results. Soil can be made fertile as deep as it can be made aerated, for, if organic matter is put down to a greater depth than the usual plow line, bacteria will follow this source of food, turning it into soil fertility as long as the supply of oxygen is adequate.

In determining the quality of work done by any implement, the condition of the soil has much to do with the quality of work done. Wet, plastic soil will not respond to the slicing process of the plow, and soil so worked becomes a hard mass when dried. Fertile soil will fail to produce optimum yields if worked while wet, and the agencies of soil building are retarded or entirely stopped.

Thus the plow can be abused, the same as any other implement, but its principle is sound. The processes of nature are slow if she is left to her own devices; the plowing process speeds up the replenishment of soil nutrients that were removed by the growing plant. Any policy of soil management that fails to recognize this fact will surely result in a steady decline in crop yields and, eventually, soil depletion. Plowing under organic matter is only part of the benefit derived from turning the soil, as aeration plays a vital role in soil building as do the soil organisms.

Agricultural writers have confused most readers by assuring them that ample nitrogen can be drawn from the air by legumes. Their statements have a serious, misguiding influence, as plants do not draw nitrogen or scarcely any of the critical nutrients through the leaf. The roots of plants are equipped with the necessary mechanism for the purpose of drawing nitrogen and other nutrients from the soil; and the amount of soil air is determined by the tillage method employed. A loose, well-tilled soil, filled with organic matter, is necessarily oxygenated. This soil air is the only source of nitrogen available to the roots and soil organisms that depend on this oxygen for existence.

Soil building stops when bacterial action stops, therefore a deep aerated seedbed can be formed to include 6 to 8 inches of the topsoil. This mattress will consume and store much more moisture than any shallow subsurface. It is interesting to note that when soils are wet and the water table is near the surface, plants develop shallow roots as the result of the exclusion of oxygen. Bog plants and those that grow under submerged conditions have air spaces in their organs, which supply the roots with oxygen. It has been definitely established that oxidation or respiration is closely connected with the metabolic activities of plant roots; hence, whatever decreases oxidation by roots, decreases plant growth and, consequently, soil-building activities.

Turning the furrow is one physical gesture of goodwill toward mother earth, and since it is the culture wherein the productivity of the soil is determined, it demands our careful and serious thought.

Chapter VI.

NATURAL ECONOMIES[1]

That we have spent our soil fertility at a rapid rate need not be alarming if we employ natural processes in replacing it. This book argues for putting the organic matter back into the soil as quickly as possible to conform to nature's own way of balancing deposits against withdrawals.

Nature lends to each living thing, plants and animals alike, enough of her vital elements to complete its cycle of usefulness, and when these living things die, they must repay this loan with a small interest. This interest represents the elements gathered and stored in the plant from the air and is an amount over and above the actual amount taken from the soil. Therefore, in undisturbed nature, each generation of plants and animals adds to the glory of mother earth. From this fund must also come that which is stolen from her by the twin gangsters, wind and water.

The uncanny skill with which nature manages her accounts can be determined from the brief time allowed between burial and renewal of these loans. It must be thus, owing to the limits of the critical elements that are available. Nothing must remain idle or locked up. No one plant or animal is allowed more than the amount necessary to complete the task expected of it. Undisturbed nature preserves the natural balance between food and feeder and, if man disturbs this balance, he will be the first victim, just as some plants and animals seem to dominate the rest but can do so just as long as the general balance is maintained.

Plowing sod to grow food breaks this natural balance, yet when we conform to nature's most vital law of quick return to the earth of all residues and hold in check the bandit we call erosion, we are making a just compensation. By contrast, it seems obvious that subsurface tillage, leaving much of the organic material useless and on the surface, will do nothing more than disturb the natural balance. Nature will call the perpetrators of subsurface tillage to account by imposing the usual penalty: declining yields.

In the business of farming, these laws are of greatest importance, for natural processes are nothing more than nature's working out and putting into effect her own laws. We farmers have observed her at work and have marveled at her precision, never deviating from the natural code. We have prized our reward for courting nature and have suffered the penalty for transgressing laws.

I have read volumes by noted writers and others, regarding man, (the farmer) and his predatory habits. So convincing was the tale of woe that I began to look upon myself as the progeny of a mountain goat or a glorified ape. Then quite suddenly I became aware that these animals would not, or could not, farm for thirty years, raise a decent family and contribute something, even though a mere trifle, to the permanence of our civilization and culture. No, in this case, if there is a goat involved, the scent emanates from the direction of the writer of such stuff.

In truth, the incorporation of organic matter is the only substitute we can offer for natural processes, and it has proven satisfactory The improved physical structure of the soil is only one advantage of organic material, for its presence is necessary if chemical action is to function properly. Analyses show that colloidal organic and inorganic soil material are intimately associated, indicating a very close relationship between humus compounds and chemical soil properties like base-exchange and the property of humus materials in acting as protective coating to mineral soil particles. In reference to plant nutrition, organic materials act directly and indirectly, for it is known that sugars can be assimilated by the higher plants. The vitamin-like particles of soil organic matter are essential to the life and reproduction of the green plants and, in turn, the source of this vitamin-like substance is, itself, the green plant. Until more is known of the necessity of the trace mineral elements and the method of replacing them to worn soils, the greater is the necessity of turning all vegetable growth back to the soil as soon as possible, for each plant has its nutritional requirements, and, if plowed under, decomposition allows those requisites to be met.

A speedy process of soil recovery could be accomplished by green manuring if a cover crop was sown to retard the loss of soil by erosion. The plowing under of green crops is well-founded for the very reason that raw plant materials possess the property of base-exchange and consequently increase the supply of available nutrients.

The sharecropper's dire plight could be altered to one of beauty and plenty if certain natural laws were practiced instead of depending on chemical fertilizers. In regions where the ground never freezes, soil agencies are continually at work decomposing the organic materials in the soil and, consequently, the need for this material is greater. The wet soil of hot climates differs from the prairie soil to the north. In these hotter, wetter geographies, the soaking process dissolves and removes the soil minerals rapidly and the excessive heat causes much of the soil gases to be drawn up from the soil and carried away. Means for trapping fertility in these regions and holding it in the soil present problems peculiar to these climates; critical becomes timely application of fertilizing material so that it will be present in the soil and in such a state as to be of the greatest benefit to the crop. In summary, soil building can be speeded up to balance with the withdrawals if the methods used conform with natural processes.

It has long been observed that animal health has been closely associated with the soil, so it is interesting to note the migratory habits of the early buffalo and wild horse herds. These large-boned animals needed great amounts of calcium and phosphorus to build their huge frames; the forage of the Great Plains region was rich in these elements. These animals never came east of these plains, even though the lush bluegrass regions to the east would seem to have been better forage, and there was no natural barrier. This gesture on the part of so-called dumb animals should have been viewed with greater alacrity on the part of the early settlers, as the bone and body-building material is essentially the same in all species of the animal world. If the forage in certain regions was shunned by the grazing animals, it should be considered a criterion of the soil's ability to provide human food.

The medical division of the Selective Service seems to bear out the fact that the health-belt of our country centers around the calcareous rich regions of the Plains, for the highest percent of acceptable registrants came from this area. The modern method of feeding farm animals is doing much to awaken scientists to the fact that more inquiry should be made into the problem of the relationship between soil and human health. The underlying reason is chiefly the fact that the human diet contains a greater variety of foods from a wide range of soils, and in this variety there is the chance of getting food from fertile as well as infertile soils. Modern science is quietly carrying out extensive research projects to formulate definite laws connecting human health and soil fertility. Until these laws are regarded as official and well established as facts, nothing will be offered to the reading public. But until such time comes, this work and others on the same subject can serve as a sample or unofficial vanguard.

There is no activity that carries greater potentiality for the human race than research on the interdependency of man and soil. When we realize that we must feed the soil the critical minerals that are gradually wearing out instead of trying to assimilate these minerals directly, then we will have started in the right direction. For example, it is known that Vitamin A is not available to the body without carotene, and carotene is associated with the green chlorophyll of plants. The healthful interplay of Vitamin A and carotene suggests that other vitamins have a counterpart that is necessary to be consumed along with them before they can perform their nutritious function. So long as nature holds these secrets, and until science unfolds these mysteries, it is folly to look to any source of life other than the soil. But it is certain that soil fertility and health are conjunct factors in the scheme of life and that the farmers of the future will be even more important to national health than medical men.

Those who fail to observe natural processes thereby fail to capitalize on natural law. For example, every day enough energy from the sun strikes the earth to turn all the wheels of industry and supply the force that puts life in the green leaf. This statement is not a literary metaphor but is actual in its interpretation. Energy is wealth, and to capture it and turn it to advantage is putting into use the natural laws governing life and growth. The farmer knows that without sun and air he could not raise crops, and yet he is not taking full advantage of these forces. Soil aeration has been discussed in another chapter, but, to capture more energy to grow more abundant crops, it is necessary to keep the soil more oxygen-infused and the crop plants in such order as to trap more sun. Since wealth is chiefly a proposition of controlling matter and energy, the farmer can, by a little extra care of the soil, capitalize on this fact and turn this vast storehouse of energy into greater yields. Clearly, the best way to relieve poverty is to cultivate such wealth.

The recent war-inspired boost in production calling for more cultivated acres has given rise to a greater use of chemical nitrogen as a soil stimulant. Even agricultural colleges endorse the use of nitrates and other heavy nitrogen carriers as a side dressing or advocate direct application to grain crops. I am not excited about reported increased yields of grain as the result of heavy applications of nitrates or sulfates as nitrogen carriers, because a long life's experience watching soil and plant reaction has proven that direct application of nitrogen to grain crops to increase yields is purely imaginary. My statement may appear paradoxical to many agricultural educators, but to prove me in error they have to resort to artificial means: increasing yields on depleted soils by the use of nitrates.

This controversy must not be left as an enigma for our returning service men, and them I will warn there is no shortcut in soil building. Many thrilling stories come from nitrogen enthusiasts who really believe they have reclaimed worn-out soil by heavy applications of this element. They tell of bumper crops of corn the second year after treatment and give the credit to the use of nitrogen. But, on being queried, they casually admit to routine plowing down of legumes or rye. What they fail to understand is that their green manure crops supplied organic matter that, through bacterial change, put organic nitrogen into the soil and introduced organic material that aided aeration. This organic material is, truth be told, directly responsible for bumper crops, and, if a farmer thinks he can duplicate yields synthetically, he is fooling no one save himself.

To hear farm specialists advocate the use of nitrogen to row crops is as disgusting as listening to vitamin advertising on the radio, for the only true vitamins are prepared by nature. It is amusing to hear the wild claims of some who stick their thumbs in their vest and fondly imagine they have solved the riddle of what makes a big ear of corn. If I were to hazard a guess, I would give all the credit to the teeming universe of soil bacteria, for surely without its help our soil would be barren. Knowing full well that these soil-builders cannot live on chemicals, I base my soil religion on my own version of the golden rule as it applies to agronomy. Feed the soil bacteria and they in turn will feed me.

Just why this fact has not been given proper place in soil management hints at the lack of study of natural processes employed by mother nature in building and maintaining our soil. I concede that nitrogen is the chief constituent of our proteins, but the maker of the universe made ample provisions for an inexhaustible supply of this critical element by shrouding the earth with atmosphere that is three-fourths nitrogen. No such lavish provisions were made for the other major elements, especially phosphorus, thus this element must be supplied to soils that show a need for it. And on soils that have been reasonably well-rotated, the need for phosphorus is greater than the need for nitrogen.

The original supply of soil phosphorus was not large and, in most cases, was the first element to need replacing. Since it is classed as an acid-radical, it does not enter into soil base-exchange reactions. But phosphorus is easily fixed in the soil unless the particular soil is strongly acid. In reality, acid reaction in soil is as important in soil building and crop growth as heavy acids are important in industry. Imagine, if you can, our predicament if sulfuric acid were denied our industries. We could not mine and process steel and its products; all industries would suffer or even cease. While not readily recognized, soil acids are equally necessary if exchange processes are to function, for this acid reaction is the foremost natural process.

Obviously, our agricultural schools are putting more stress on economic problems than the intimate problems of the soil. This might be expected and even encouraged because the average farmer finds these problems outside the sphere of his everyday experiences. Small wonder, then, that titled educators take a special interest in this branch of farm business, for farm economics is the brainchild of our national planners, and their prognostications can be dispensed from the round table with dignity.

Not so with problems of the soil. To master its meaning is to live on it, live by it, ever watchful for any deviation in response due to variations in tillage, getting the feel of it in the spring and working and observing it throughout the changing seasons. There are too many articulate people whose inspirations are prompted by the glamorous thoughts of nature being the mother of us all. They speak and write pretty things to fire our imaginations and leave us with a literary hangover. The present-day farmer wants facts, not fancy, something workable that will help analyze natural processes that build our soil. To these processes we must look for our future existence.

Chapter VII.

MAN AND NATURALIST[1]

It is not surprising that we farmers should question the findings of early scientific writers in light of advancements in agricultural science. A noteworthy example is the work of Darwin, the foremost naturalist of all time. Darwin was evidently not a man of the soil, for he failed to connect the changing proportions of the plants and animals to the changing soil.

It would seem to me that to interpret the natural laws of plants and animals without proper cognizance of the part soil played in their evolution is but inviting controversy. Humans would have had the same fate as the lower animals were it not that they were given the power of thought, for as their food supply faltered, both in quality and quantity, they moved to virgin soil where new life in the soil put life into their food. Plants and some animals, however, accepted their fate in the only manner possible: slow degeneration. The animals with the greatest range of activity maintained the characteristics of their species for a longer time, but time was the eventual winner. Our present domestic animals would share the same fate and degenerate even more rapidly were it not for our knowledge of soil science, and we too would have to pull stakes and hunt virgin soil.

In fact, ongoing principles of evolution might be better expressed in terms of the gradual disappearance of vital soil elements. Darwin failed to link changes in the species to the changing diet, and millions of years would necessarily use up many vital forces necessary to the well-being and development of the animal population. If he had taken these facts into consideration, the principles of his work would have enjoyed still greater applicability.

Evolution is necessarily a forward movement. Each species must glorify its predecessor and so on until, as some claim, small animals assume huge proportions, and reptiles assume the proportions of man. These scenarios are sobering at first thought, but seem less feasible, and therefore less threatening, when the evidence is considered. Evolution must give way to the very simple fact that everything that has life is of the earth, air, and water. A farmer, perhaps better than any other worker, knows things of the earth cannot reach greater proportions than their species. Man's vanity, then, is the prompting force behind his refusal to accept the simple equations of life. He befuddles his mind, strains his imagination, and tests his ingenuity to make these equations seem complicated and important. But the final deductions are simple and are expressed in the language we all can understand; everything depends on food; food depends on soil, and soil depends on its management.

Darwin succeeded in convincing scientific men that the law of evolution is true of all living things. But it must be considered that early scientific men had but little knowledge of the relationship between soils and the plant and animal world; if they had, Darwin's conclusions would have been greatly modified. To name an instance close to the farmer, crop rotation points out that strong foragers of certain mineral nutrients do not follow themselves in the fields year after year without serious consequences; the continuation of the same crop will finally lead to barrenness. Nature's own process of rotation is similarly slow, and when native plants use up soil elements critical in their development, they depend on natural means to spread to new soil. Only after they migrate to hostile climates do they risk extinction. In sum, as long as our topsoil is kept fertile, we need not fear any unwanted variation of the species.

Chapter VIII.

HIDDEN HUNGERS[1]

We are what we eat is becoming an old chestnut. Though this statement is quite true, it is also wrongly sending many people to the corner drugstore for their quick-fix daily health doses when the real problem lies in our depleted soils.

Hidden hungers amount to a slow starvation for the critical soil elements that are gradually wearing from our soil. Fortunately, this deficiency can be detected in both the plant and the soil, and practical men such as the University of Missouri's Dr. W.A. Albrecht and others capable of interpreting natural processes can be relied upon for sensible conclusions. Today's farmers owe much to Dr. Albrecht, as his appraisal of hidden hungers in our soil and food plants has spurred interest in better pasture and meadow management. Farmers and ranchers are now quicker to determine the net results of pasture and meadow improvement practices, for in them lie the possibilities of improved animal health with its accompanying higher financial return.

The green leaf is nature's factory and the medium of exchange between man and nature. In this dross of earth are the essential chemicals that go to make up the formula that has baffled scientific men the world over and is regarded as the key governing life. When the natural soil elements are present, we can assume that nature, as any good chemist, will perform her function with uncanny precision, giving vigor to the plant and health to man and beast.

These hidden hungers originate in the soil and reach us by way of plants, who also want for proper nutrition. Animals also suffer hidden hungers, and humans consuming the products of starved plants and animals also suffer. The whole series of torments are caused by nutrient shortages in the soil. Proper nutrition is an enemy of disease, in plants as in people. Fungus attacks on plants result in the "damping off" disease, which is related to a hidden hunger for calcium. More recently, potato scab has been connected with insufficient calcium in relation to potassium in the soil fertility offered the potato plant. The potato plant demands much calcium in its tops, which duplicate red clover in content of this nutrient.

Plant health is doubtless declining as the soil is declining. As we stewards fail to return manures and wearing soil mineral nutrients, we are invoking hidden hungers in the plant and encouraging plant diseases. Why can we not always see these hidden hungers? Man's transitory habits have obscured much of what a life in a limited geography might otherwise reveal. The recent shipping blockade as well as rubber and gasoline rationing have put a fence about us, as we do our livestock. Our own deficiencies or hidden hungers ought to lead us more quickly to considerations of the soil. Calcium and phosphorus deficiencies, for example, soon show up in livestock. In some localities animals born in the early winter develop rickets by late winter or early spring. Their bones break readily and the pelvic-spine joints separate. Disturbed reproduction processes are another consequence of hidden hungers for calcium and phosphorus. Shy breeders among farm animals are on the increase, and this condition is easily traced to soil deficiencies. In some of the less fertile regions, farmers are wondering why their cows breed only in alternate years. Is it not suggestive that on many of these same soils the calves show malformations that make them less true to breed types?

The backbone of a cow reared in such a herd reveals she had suffered deficiencies and had sacrificed part of her backbone for her fetus. Naturally, she used the succeeding year to replace her backbone rather than indulge in another cycle of reproduction. Cattle that have been regular breeders have become breach-prone and subject to premature death. Males, too, may lose breeding capacity on inadequate forage feeds grown on deficient soils. Male rabbits, under experiment with forage grown on calcium-deficient soils, become impotent, while their littermates, on feeds grown on calcium-rich soils, retained their capacity to father offspring. When the feeds of these two lots of rabbits were interchanged, the situation, not surprisingly, was reversed. Those originally made impotent by lime-deficient forage recovered their male potency, while the lot put on forage from lime-deficient soil, lost it.

Reproduction is a delicate physiological mechanism—its late arrival and early departure in the individual life cycle serve as a litmus test for unseen soil deficiencies. Hidden hungers, by way of decreasing fertility in the soil, may be the quiet force by which species have become extinct. Nature's warnings allow time for us to heed them against our own extinction if we will look to the soil and conserve its fertility against the forces of exhaustion as well as its body against the forces of erosion.

One might say that the moods of mother earth are not always constant and kind toward those creatures whom she nurses. The rates of delivery to the plants of exchangeable nutrients by the soil vary with different seasons. Because plant growth stops during the winter season, this need not imply that chemical reaction within the soil ceases during the same period. In fact, the rapid exhaustion of the exchangeable supply by accelerated plant growth should occasion restoration of equilibrium with cessation of plant growth. In the absence of plant growth, nutrients may ready themselves for future absorption during the next growing season; spring plowing facilitates these adjustments of concentration of available nutrients by rearranging contacts between colloids and mineral crystals. Consequently, the spring plant grows into this larger supply of fertility, and through root extension for its rapid absorption, meets a liberal supply of soil nutrients in contrast to the lower rate at which the plant is building its carbohydrates. Such plants are mineral-rich and protein-rich, to say nothing of their contents of vitamins or other growth-supporting essentials. As the plant growth extends into the summer season, the store of exchangeable nutrients diminishes. At the same time, the increase in sunshine and in temperature brings greater rate of carbohydrate production. Thus, spring growth of plants means high mineral and protein concentration of nutritive value as body construction, but later growth in the summer suggests woodiness and fuel value.

When, with each mouthful of luscious grass, the animal ingests growth-promoting nutrients assembled by extensive root action through the soil—rather than principally from a plant top collecting sunshine and fresh air—is it any wonder that the shaggy winter coat of hair is shed, and the animal is sleek by May? Should there be any wonder that animal thriftiness is lessened by August, and shall we accept the oft-given explanation that it is wholly because of the heat and bothersome insects? Some of you have heard cattle feeders say of steers in the feedlot in February that they ought to go to market soon because they are licking themselves. Some even say that this licking behavior is an index of feeding efficiency. It would be perhaps better to view it as a danger sign that recommends sale before disaster. Have you ever asked yourself why cattle do not continue to lick themselves, or each other, after they have been on fertile spring pasture? Might it not be possible, after a long period of winter in a dry lot and on dry feeds of low mineral content, that the body's need for Vitamin D becomes greater than the ingested supply? Might not the animal's fatty secretions by the skin become activated in the sunshine with sufficient resulting equivalent of Vitamin D, which the animal by chance has learned to keep going through the cycle of excretion, activation, and ingestion?

In support of this theory, it has been demonstrated that the yolk of the sheep fleece for the fat of the wool is more prominent on sheep fed hays grown on land limed and phosphated than on those fed hays grown on lime-lacking soil. Sheep growth per unit of feed consumed was also better by from 25 to 50 percent on the former. The better body growth and more prominent skin secretions may both be connected with the same physiological improvements linked to soil treatments. Soil fertility is not flowing into the plant at constant rates at all periods of the season. These differences through the year and in variable soils make for different plant compositions but exert still larger influences on grazing animals.

Naturally, the question implied is this: Do humans escape these subtle forces of the soil? That seasonal differences in soil fertility exert themselves on animals is not so difficult to understand when we recall that nature has put the birth time of animals so commonly in the spring, or the time when the heavy calcium load of lactation can be met more widely by the luscious vegetation for the survival of the species.

In effect, unless the soil supports the vegetation, the vegetation cannot support the animals. Human births are not seasonally concentrated. Yet the December 1941 issue of *Human Biology* reports from a survey of ten thousand students at the University of Cincinnati that those born in the spring are taller, heavier, and smarter than those born during the summer. Here may be further evidence to the more direct relation of soil fertility to the animal and human species. Here is possible human evidence that mother nature nourishes more efficiently in the spring than in the winter. Experienced farmers now know that better forage and improved soil fertility can help our maternal animals avoid hidden hungers and better carry the reproductive load of fetus development.

Leguminous crops have been one of the great feeds to bring breeding animals through the winter in good shape and health and with a generous crop of offspring. They have been fine means of growing young animals with good bone and straight backs. Farmers all want plenty of legume hays but cannot grow them, they say, because their soils are sour. Legumes need lime on the soil to remove acidity, the common enemy of legumes and especially alfalfa. It has only recently been demonstrated that the soil acidity is not injurious of itself. So-called plant injury by soil acidity is largely a matter of a deficiency of the plant nutrient calcium. It is, in truth, a deficiency of many items in soil fertility of which calcium is chief. We have believed that liming gives benefits because the carbonate of calcium removes acidity. In reality, benefits come about because the calcium carbonate puts the calcium into the soil to satisfy the plant's needs for this nutrient in its bodybuilding requirements. After we have come to understand the function of liming, we shall be putting calcium back to the soil and driving out some of the hidden hungers in our livestock. When we are once fully accustomed to habit of introducing calcium, we shall be more ready to do the same for phosphorus and all the elements that we will eventually put back to relieve the animals and ourselves of all these hungers. When only a few hundred pounds of limestone, less than that of phosphate, and still less than tens of pounds of the other nutrients, are all that must be put back to make normally healthy plants and animals, we should shake off our pessimistic views, dispel our fears of the hidden hungers, and go to work optimistically in restoring soil fertility.

Soil acidity has so commonly been considered disastrous that perhaps you may not be ready to believe that the soil acidity so prominent in the temperate regions may, in reality, be a blessing. The highest degrees of soil acidity occur in regions of moderate temperature and moderate rainfalls—where the maximum concentration of human population occurs. We have been inclined to believe that disturbed body comfort relative to temperature works against denser populations in the frigid or the torrid zones. But when human movements, like an army, are dictated more by food availability than by bodily comfort, we must look to their increased chances to be fed there as the cause of concentration of people in the temperate zones.

Soil acidity represents these greater possibilities. The very property of clay that enables it to hold much hydrogen to make it acid is the same property that enables it to hold large quantities of many kinds of the nutrients of plant production. Soils of the frigid zones do not develop much clay content, or a clay that is able to supply much plant nourishment. Soils of the humid tropics, by contrast, have a clay product formed from such complete mineral breakdown that the resulting clay compounds have little holding, or little exchange capacity. So the acidic property of the clay, and of its associated humus, formed in the temperate zone may hold the reason for our settlement patterns as a species. High concentrations of people and the food to feed them have been supported where soil acidity mobilizes the nutrients into the plants more effectively than under soil neutrality. Recent studies using spinach as the test crop have demonstrated that this vegetable took more calcium, more magnesium, and others from the soil when these exchangeable nutrients were accompanied by acidity.

We must not console ourselves too quickly, however, with the belief that soil acidity is always beneficial. It is beneficial only when accompanied by nutrients. As these nutrients become exhausted, hydrogen or acidity replaces them, and when acidity is all that remains, the plants must suffer starvation. Such was the situation in our virgin soils. They were acid but fertile because of high humus contents in which fertility had been hoarded by virgin vegetation. These stores have been mined. Over time, as crop yields declined, the blame for failed legume crops was increasingly put on soil acidity. However, it was not the presence of acidity, but the absence of fertility, that was the cause. We now know that nutrients in fertilizers put into the soil are taken by plants more effectively in the presence than in the absence of acidity. The crusaders against soil acidity would do well to join our cause instead and march under the banner of soil conservation.

That some of the hidden hungers in animals and in plants can be driven away by putting calcium and phosphorus on the soil as fertilizers has been experimentally validated only recently. Sheep, as test animals fed the same amount of hay and grain per head per day, gained weight variously as the soil had been given phosphate, or lime and phosphate. The plants had some of their hidden hungers routed, too, as judged by the different yields per acre. The internal physiology of the plants must also have been changed since they were not widely different in chemical analysis, yet served widely differently in the efficiency with which their offerings in bulk served to be converted into meat as mutton. Viewed more specifically, if the sheep gains are calculated as coming entirely from 2 pounds of hay taken a day, the bean hay from untreated soil made 1 pound of gain from 9.7 pounds of hay consumed; that soil given phosphate produced a hay of which 5.4 pounds made one pound of sheep. At last, when the soil was given lime and phosphate as a means of more properly satisfying the soybean plant hunger, then 4.4 pounds of this hay was all that was needed to give 1 pound of lamb in return. The studies tell us this: even if we should not be moved to change our practices out of sympathy for the animals that are suffering hidden hungers, we should now feel compelled by economic imperatives.

Like the proverbial canary in the coalmine, rabbits have been used extensively as forage eaters to test the hay crop for its dangers as an inducer of hidden hungers, or conversely its efficiency as a producer of healthy animals. This miniature form of livestock is serving as excellent biological assay of the soil's store of plant nutrients, calcium and phosphorus, in their mobilization in the body from the intestine to the bloodstream to the bones. Likewise, animal and human hungers exist for energy-supplying compounds as they do for the growth-promoting substances, such as protein that contains nitrogen and phosphorus, and for calcium and other elements going into the plant from the soil. Thus when soil-given nutrients are denied the animal, it may be even possible that the animals are not using their energy foods efficiently. There are suggestions that lime and phosphate on the soil may play an important role in aiding metabolism, the chief process in building new tissue in animals and humans alike. For example, close observation of grazing animals suggests their degree of discrimination in selecting herbage according to soil treatment. Hogs and cattle, we know, will choose certain parts of the pasture to graze and, in many cases, will graze the forage close to the ground while other portions of the pasture or meadow stand untouched.

Our burgeoning awareness of the role of lime and phosphate in preventing hidden hungers suggests the possible shortage of other nutrients that are, at present, considered merely minor elements. If their addition to the soil can set in motion a natural means of nutrition—one that works from the ground up—the needless searching for drugstore dosages as antidotes against hungers, bad health, and impending death will become obsolete. We need to see in the restoration of the soil a means of demonstrating the age-old truth that an ounce of knowledge-as-prevention is worth libraries of information-as-cure. Conservation of soil fertility needs to be more than ideas. It must be action out on the soil. The metabolism in our soil has been overstimulated, and it must have periodic rest under the sod, for just what goes on under this mystic blanket is out of this world.

As we get a fuller knowledge about the soil and the role it plays in making mother nature a better nurse of us all, our present prodigality toward soil will change to a reverent conservation. And if we but put into practical operation the knowledge we have gleaned from nature and the kindly exchange of thoughts and experiences one with the other, then our lives will be more abundantly blessed and our valleys more pleasant.

Chapter IX.

THE OLD QUAKER WAS RIGHT[1]

After making an extensive study of soils and plant relationship, I was determined to satisfy myself in the matter of just how much corn could be produced on an acre of ground with all conditions under human control being favorable. My experiment was spurred on by reports of fabulous yields coming in each year from various corn yield contests sponsored by county Farm Bureau units and seed corn companies.

I had experimented each year for many years, in a small way, with plants and soils, sometimes marking a few hills of corn using combinations of fertilizers to determine their value; using varying amounts of lime to establish pH preference; comparing deep and shallow seedbeds in respect to the water-holding ability; evaluating the advantages of a dust mulch and determining the maximum amount of organic matter the seedbed could utilize. These seemingly insignificant experiments supplied facts that cannot be easily erased from my mind and the results of these experiences, coupled with a strong belief that the traditions of the furrow form an integral part of our advanced agriculture, were the motivating force for this work.

I had always wondered just what limiting factor would prove to be chief in determining the ultimate yield. Accordingly, in the summer of 1943, I selected a field that was favorable in topography for either a wet or dry season. This field happened to be in clover, and I started to build it up with sufficient soil nutrients to support a maximum possible yield. The pH value of this soil was the exact preference of corn, and the other major minerals were supplied with both commercial and barnyard manures. A heavy application of manure was spread on the lush clover and promptly plowed under. Although I regretted plowing this ground in the fall, I reasoned that I would spread another heavy coat of manure on the surface of the plowing to compensate for the loss of the cover crop through the long period before planting time. This was done immediately after the ground was plowed, and nothing more was done until May first, the following spring. At this time 400 pounds of 2-12-6 commercial fertilizer were applied per acre. This application of fertilizer was to insure plentiful amounts of these nutrients in case they were needed to support the heavy stand I expected to plant.

After all, the thing I expected to prove was something that has never before been satisfactorily established: synthesis. I wanted to know just how much leaf exposure to the sun was necessary to perform the vital function of manufacturing the protein, carbohydrates, and fats that go to make up our food. It is generally known that the chlorophyll carries the raw nutrients to the leaves, where the penetrating rays of the sun turn the nutrients that have been gathered by the roots in the form of soil solutions into the crude proteins, carbohydrates, and fats that are later fed to the grain head or corn ear.

In order to arrive at worthwhile conclusions, the planting rate started with the normal stand of 10,000 stalks per acre and against this plot the thicker stands were checked. The second plot was planted at the rate of 15,000 stalks per acre, the third at the rate of 20,000 stalks per acre, the fourth at 25,000 stalks, and the fifth at 30,000 stalks per acre.

The 1944 growing season was typical of the Corn Belt. The growing year began warm and humid, inducing rapid crop growth, giving way to dry days in late July and early August that, in turn, ended with good rains, as usual, enjoyed before any great damage was done.

The plots of 30,000 and 25,000 stalks per acre produced such a thick mat of foliage that the direct rays of the sun did not fall on the ground, thus eliminating much of the natural evaporation. This fact seemed to compensate to a great degree for the extra moisture required by the heavy stand. During the peak of the dry period, the soil in the heavy stands seemed to be more moist and aerated than the thinner stands, although definite hunger signs were beginning to register. Since the soil was loaded with an abundance of necessary nutrients, I suspected that this hunger was for sun. The thicker stands permitted only a few of the top leaves to come in direct contact with the sun, and while the stalks attained their normal height, their diameter was greatly reduced, and the lower leaves proved pale and unthrifty. However, the size and quality of the ears and eventual yield told the story of synthesis and satisfied my curiosity.

As was suspected, the normal stand produced large, sound ears of fine quality corn, checking out at 130 bushels per acre. The next plot with 15,000 stalks showed only a slight decrease in yield, but the quality was not so good. The yield of this plot was 127 bushels per acre. The plot with 20,000 stalks showed a decided reduction in yield or 104 bushels per acre. As the ears diminished in size, there were more round, black grains, giving the corn a low-grade appearance. The plot with 25,000 stalks yielded but 88 bushels, and the last plot supporting 30,000 stalks trailed with a scant 70 bushels per acre. This experiment, though crude, shows the value of sun in the productivity of our crop plants, and further suggests that planting should be done in such a way as to allow a maximum of leaf exposure.

In all fairness, however, this difference in yield of from 70 bushels per acre to 130 bushels per acre cannot be attributed solely to the lack of sun. Somewhere in the secret archives of nature's code is a law that has not yet been formulated by man, but in effect controls the amounts of soil nutrients that can be released annually. This is only a logical conclusion of my many years of observing the behavior of crops and soils. The excessive plant numbers on a unit of land promote strong competition and ultimately result in impaired growth. These facts are being reflected in the planting rate and yield of soybeans. The old method of broadcasting or drilling at the rate of 2 bushels per acre is giving way to row cropping using but 1 bushel of seed per acre. The beans planted in rows grow taller, and this longer stalk or stem supports many more pods and consequently makes greater yields.

While my costly experiment with corn yield did not break any world records, it did prove to me something far more worthy—synthesis is one of the major limiting factors in crop yields. It is yet another natural force that the old Quaker must have encoded in his warning to me as a small boy: *If you try to beat nature, nature will beat you.*

Part Two

Soil

Chapter X.

SOIL AND TILTH: A PRIMER[1]

Some mistakenly hold the idea that soils were created perfectly balanced and that they became depleted due to mismanagement. In fact, only a small percent of the earth's surface is suitable, or ever was, for agriculture. And even in the best producing areas there is a wide range of soil adaptability. In general, soils are regarded as quite simple, but they are actually very complex, both in a physical and chemical sense. Soils, indeed, have a three-fold nature—physical, chemical, and biological.

The bulk of our soil was formed by eroded rocks, clay, sand, gravel, and organic matter. A clay-like material forms the ideal subsoil, while a portion of clay and fine silt makes excellent colloidal material that holds water and collects fertility. The organic matter or humus is largely in the top 3 inches of the soil structure giving the soil its dark color. Most of this matter is derived from plants and is present in different degrees of decomposition. The vital necessity of organic matter cannot be stressed too much, as it not only adds to the general tilth, but contains large amounts of nitrogen together with other elements of plant food that is made available by bacterial change.

The deep, well-directed furrow is the one factor that is widening the range of the soil agencies responsible for the building of a more fertile soil, and, instead of a thin layer of fertile soil, tilth can be accomplished to a much greater depth, thus storing more fertility and moisture. In soils of sufficient organic matter, bacteria start working as soon as temperature and moisture conditions permit. The soil with a maximum amount of humus will naturally culture more bacteria, hence, fertility will be more promptly assured.

There are several groups of bacteria in the soil, and while some prove harmful to plants, their action can be largely controlled by keeping the soil free from acids associated with those particular groups. Therefore, the first step in soil building is to determine the pH value of the soil—that is, whether or not the soil is acid and in need of lime.

Lime is not classed as a fertilizer, but is known to be the most important basic element, since it is readily acted upon by water, and hence becomes a basic part in soil solutions. Since it combines with the chief soil elements, it is the principal performer in base-exchange. Without lime, the fertilizing qualities of nitrogen, phosphorus, and potash cannot function properly.

Magnesium is regarded as second to lime in building soil fertility, but since it is a constituent of lime, its presence in the soil can be determined by the availability of lime. In chief farming areas, the soil was originally balanced with the critical soil elements in amounts to last many years without replacing. This is particularly true of the trace elements. The major elements—lime, phosphorus, and potash—are consumed in great amounts by the growing plant and hence become a matter of concern. The most important element, nitrogen, has ceased to become a bottleneck in crop production with the purposeful cultivation of legumes.

The original supply of the heavy elements has lasted through years when transportation of heavy material would have been prohibitive. Lime reserves have been strategically placed and have served their purpose well. Phosphorus, however, must be moved great distances, and had the soil been originally deficient in this critical element, our national progress would not have been so rapid. Other areas, however, seem to serve as stockpiles for the necessary elements, and while these areas are unsuited for agricultural purposes, they serve agriculture's needs for these materials.

Providence was kind when it placed nitrogen in inexhaustible amounts in the air, but it left the process of capturing and turning it into foods to the devices of the tillers of the soil. Oxygen is best thought of as a spark, and this source of life can be breathed into the soil by employing the proper tillage methods. We farmers know that the greatest source of free nitrogen is in the air; therefore it is only common sense to make the soil more aerated if we are to build and store fertility in the soil.

Natural order in soils is based on the changes produced by surface and subsurface deposits of loose geologic materials that, over time, have developed distinct characteristics. According to the law of soil genesis, the same natural forces acting under like conditions on deposits of loose geologic materials produce identical characteristics. This law explains why many soils widely distributed are identical. Certain masses of earthy material deposited in arid climates will change their general character to conform to desert standards, and, likewise, the same materials placed in humid regions will eventually change according to climatic conditions.

Soils, then, are classified according to their similarity, and while there are many individual soil types, a few common classes are only of general interest. Names such as Tama silt or Chester loam are used locally to distinguish a certain type of soil. But class names such as Miami silt, Dunkirk, and

Chernozem are quite generally used to designate large classes of soils. Too many look on these names as soil types, but in reality each name represents a great many different types of soil that share a basic likeness.

The Russian scientist Dokuchaev believed that if we were to consider carefully the factors concerned in the formation of the soils of a certain region—such as vegetation, precipitation and temperature—we should be able to tell in advance the nature of the soils we might find there. He no doubt had in mind the law of soil genesis that is generally recognized as a basic law governing soil classes.

Soil relationship may be interregional and embody the same characteristics the world over, such as Chernozem, Dunkirk series, and podzol. This does not mean that soil of the Chernozem class found in Russia is identical in most of its physical properties with the same class found in the United States. It is identical in most of its properties, but may not have the same productive power, since fertility is a quality of the soil and not a part of it. As there are so many factors entering into a soil's ability to produce, that fertility cannot be included in soil classification.

On a mathematical basis, our common silt loam soil may be 35 percent sand, 50 percent silt, and 15 percent clay. A family group of soils may be divided into two subgroups. One subgroup may represent soils that have been formed from the same kind of parent material, but which accumulated in a different manner—material on a steep slope developed from a residual geologic matter and transported as alluvium. The fertile and highly productive prairie soil was developed under the cover of lush prairie grass, and the protecting roots of this grass conserved the soil nutrients. While great amounts of organic matter have given the prairie soil its rich, dark color, there are many soil types and colors in the comparatively small area comprising this class of soil.

Hydromorphic soils embrace many soil types developed under extreme wet conditions or permanent saturation. Many names are used to designate the divisions of this class, including peat, bog, fen, moor, marsh, swamp, wet podzol, and tundra. Most of these soils are acid and will grow only such plants native to them.

Since soils are always in the making, it becomes apparent that the biotic soil-forming agents, plants, and animals must be considered for the prominent part they play in soil building. Plants are more prominent in the natural soil-building scheme than the other agents. When one considers that soil fertility and profitable crop production hinge on vegetable growth that is returned promptly to the soil, the problem of productivity need not seem so confusing.

The fertile prairie region of our country has the largest long grass prairie soils in the world, and, coupled with favorable climatic conditions, this soil stored up a vast amount of almost inexhaustible fertility. Nature accomplished this marvelous soil-building job without any interference, and to some it might appear that this soil is wearing out due to the predatory habits of the farmers. But comparative yields of crops should dispel these fears, since crop yields have reached climatic levels that are the supposed limit on any soil.

Soil Productivity

The earth's population has never been well-nourished, and many human ills, running to epidemic proportions, have been the outgrowth of an insufficient amount of protein in the diet. Yet we continue to waste the productivity of our soil.

We cannot increase the production of the critical protein unless we cease depleting the soil of the constituents that go into the production of this all-important food. In the past we unconsciously followed a rotation program that was responsible for maintaining productivity, which embodied a three-year rotation of corn, oats, and clover. This program kept the organic matter and organic nitrogen at proper levels, and if proper significance had been given lime and phosphorus, the problem of soil productivity would have been completely solved. With the present knowledge of the necessity of lime and phosphorus and their relation to the natural organic fertility especially afforded by legumes, the present-day farmer can easily maintain the productivity of his soil by practicing the three-year rotation program.

Owing to the urgent need for more vegetable oils as a critical war and reconversion material, the soybean has broken the usual rotation program on many farms and has added an extra year to the plowed-up state of the soil. If this is continued, it will mean a steady loss of soil humus and will hasten depletion. When legumes are used as the nurse crop for oats or other small grain, the legume hay can be cut the next year, and the stubble and second growth will insure a satisfactory cover crop for the coming winter. This will guard against erosion and supply a good amount of organic matter when plowed in preparation for corn. Careful tests, over a period of years, show that organic nitrogen and general fertility remain the same when the ground is tilled one year in three. The inclusion of beans in the program simply adds one year of exposure to the elements of erosion and lessens the amount of humus.

Many farmers are deceived by the fact that beans are legumes. Surely enough, they are, for they gather nitrogen and store it in nodules on their roots. But other factors enter in to the bean problem that make the bean a soil robber. Beans render the soil extremely loose, and for this very reason, bean ground is an easy mark for leaching and erosion. Taking into consideration the extra year the soil is left nude and its physical character inviting natural hazards, the continuous use of beans will surely reflect the truth of these statements on future corn yields.

The tragedy of continuous cropping and its effect on productivity is proven in a comprehensive manner in the Morrow plots at the University of Illinois. These plots have been in continuous production since 1876, and while they show the effect of continual cropping, they also show the marvelous ability of the soil to come back when given the proper care. These experiments have definitely proven that the mere replacing of the soil nutrients consumed by the crops was not enough to put the land back in its highest bracket of productivity, but other conjunct factors were also responsible for much of the gain in yields. Legumes added organic matter, thus improving the texture of the soil, allowing aeration and percolation. The benefit of close-growing crops two years out of three speeded up building processes, and all that was necessary to reclaim the depleted soil completely was the addition of lime and phosphorus.

When depletion was to such an extent that only 12 bushels of corn could be raised per acre, the protein content of this pitifully small yield was likewise low at around percent. Present-day farmers cannot afford to raise such low yields, but our knowing the penalty for continuous cropping is a valuable aid in determining our rotation program. When fertilizers were added to the depleted soil, some increase in yield was noticed and also some increase in protein and general quality. But not until the plots were put on a three-year rotation program, with the addition of lime and phosphorus, did the yields reach optimum proportions. As long as this method was practiced, the determining factor in fluctuating yields was merely the weather.

On a protein per acre basis, the Morrow plots have shown an amazing difference between corn grown from depleted soil and the product of soil where the three-year program was used. The yield of 12 bushels from the depleted soil tested less than 50 pounds of protein per acre, while the heavy yield of 100 bushels per acre on the rotated plots showed a protein yield of 640 pounds per acre. This means that 40 acres of properly managed soil will produce as much protein as 400 acres of continuously cropped soil.

In planning soil improvement, attention should be divided between cropland and permanent pasture. Since the permanent pastures are not plowed to incorporate humus rapidly in the soil, the basic mineral nutrients, such as lime, phosphorus, and nitrogen, can be spread on the surface and allowed to percolate through the soil. Pasture improvement has long been neglected, but it seems certain that more return from fertilizers can be realized when used on pastures or fields where a greater amount of foliage is desired.

If we are to understand just what makes soil productive, we must recognize the fact that soil health is brought about by soil bacteria, and since these soil agencies are living organisms and must have food, it is a matter of chief concern to place and maintain generous amounts of organic matter in the soil at a good depth. Without these organisms, no soil-building program would function and nothing would grow. Some are inclined to believe that by inoculating legume seed it is possible to increase the bacterial population of the soil. This procedure will work on soil of ample organic matter, but on worn-out soil, nature will strike a balance between organisms and their food supply. Evidence of the ability of various organisms to multiply rapidly is all around us and is the keystone for all processes that make life possible. An animal dies, and, if left alone, will soon be devoured. Similarly, manure or even a cornstalk can be plowed under and soon it will decay, which is nothing more than bacterial change. The more food for these organisms, the greater their population will be. When we overlook the fact that nature's soil-building scheme centers on soil organisms, we fail to cooperate with nature in reclaiming worn-out soil. Soils cannot be built up with chemicals alone. I have tried this shortcut many times only to find that the principles embodied in the old law of diminishing returns remain true.

Using the three-year rotation with the addition of lime and phosphate best solves the riddle of soil productivity as well as the dietary needs of man and animal. The effect of feeding the soil for the production of feed of higher protein and mineral content was first noticed by the dairy farmers. This might be suspected, since the professional dairyman keeps a close check on feed and return. Nutrients fed to the soil stimulated milk production far more than when concentrated forms of various minerals were mixed in the feed. The greatest return, however, was the ability of the crops to produce greater amounts of protein by the proper management of the soil.

Lime and the Art of Liming

Until recently, the farmer had little incentive to look into the chemical nature of the soil he tilled, since it seemed to produce without any interference. Consequently, when word began to grapevine through rural areas that the soil was collecting acids due to the depletion of calcium, it was accepted as evidence that, if optimum yields were to be maintained, the land must be treated at once. Lime-haulers were suddenly kept busy, and the pitiful part of it was that many farmers spent many hundreds of dollars that could have been put to a better use.

Why? Because the spreading rate was started by the first customers at 3 tons per acre, and on most of the soils, one-half of this amount would have been sufficient. Most of the soils in the Corn Belt show some acidity, and, if alfalfa is to be grown, it is imperative to add the limit of lime, or an amount necessary to bring the pH value of the soil to two points above neutral. I have personally tested soil that has been in cultivation since 1850 and found it to test slightly under neutral—the exact preference for corn.

Liming is an individual matter, and if the most good is to be derived from the money spent, the soil should be tested and lime applied according to individual cropping needs. Many Corn Belt soils have no need for lime, since alfalfa grows in profusion, which constitutes a rigid test for the presence of this vital element. A certain amount of acidity is necessary if the soil elements are to function properly, and there is danger of over-liming. More of our cropland is alkaline than acid, thus our experiences in handling acidity are quite elementary.

Accounts left by the Romans indicate that "marling" the ground was practiced by the Gauls as early as 500 BC The Flemish farmers, noted for advancement in agricultural practices, had no need for marl or lime until AD 1600 and knew nothing of its action until English husbandry introduced its use. As early as 1649, the English writer Blith, in discussing the question of liming, warned of over-liming as well as under-liming. At that early time, little was known of the chemical make-up of the soil, and the misuse of lime led to a wide difference of opinions as to its value. And today, too, few people know what function lime performs and few use it wisely. Lime in itself is not classed as a fertilizer but it is a major nutrient element. It is readily acted upon by water and therefore becomes an active basic element in soil solutions and the principal action in base-exchange compounds.

There are many natural sights that indicate the soil's need for lime, but the surest way to determine the amount that is needed to correct acidity is to acid test. The lime requirement of soils is the amount needed to neutralize the total acidity to a given depth. This amount may vary as much as 3 tons per acre, so, at this great disparity, it is a matter of economic importance to apply what is actually needed. Personal experience in testing the soil has revealed many interesting facts regarding acidity. Some soil that has been in cultivation for seventy-five years required but 1 ton per acre, while other nearby soils showed a need for nearly 4 tons per acre.

In cases where only 1 ton per acre is needed, it not only represents a waste of money to raise pH values too high, but a heavy application of the soil sweetener would place the soil in the alkaline category, which would be out of the preference range of grain crops. It cannot be safely assumed that lime will stay in the ground until needed by the crop, as much of it will leach away before it is used. The wise thing to do, then, is to lime more often, using just enough to correct acidity and meet the actual needs of the plant.

Soil fertility does not depend on the presence of lime carbonate, nor does it follow that soils devoid of lime carbonate are infertile. The fact is, for most crop plants, free carbonate of lime is not required. For certain crops, some soils are fertile despite the fact that they have low pH value or high lime requirements. Most plants show calcium in their ashes when burned, indicating that there is a specific need for this element in their growth. But if this were all the function of lime, many of our soils would need no extra application for ages to come. Heavy soils, or those that contain much clay, are difficult to till when they become acid since the acid clays puddle easily, becoming impermeable.

Through biological processes, carbonic, sulfuric, nitric, and humic acids are ever forming in soils, so that active hydrogen ions continuously tend to displace the cations of the base-exchange compounds. Carbonate of lime tends to neutralize these acids as they form, thereby protecting the exchangeable basic elements, so long as carbonate of lime is present. After the disappearance of added lime in a soil as carbonates, the calcium and magnesium, which have become constituents of base-exchange and other compounds, become important sources of these elements in plant nutrition.

The need for lime in some plants is imperative, and, in fact, they will not thrive on acid soil. In the case of alfalfa, tests show that half of the basic elements absorbed are bicarbonates. On the other hand, timothy utilized half as much. It is thought by some that alfalfa will build the fertility of the land. This, in part, is true when the land is properly limed and the seed inoculated. Nitrogen will be gathered from the soil air by the alfalfa and the nitrogen supply of the soil increased. But there is another side to this. Alfalfa is a heavy feeding crop and requires many other essential elements of plant food, especially phosphorus and potash. If these elements are not supplied in sufficient amounts to overcome this added strain, then alfalfa will become a soil robber.

It seems certain that the plants consuming a maximum amount of lime use it as a constituent of protein. In order to increase the value of our grain and feed crops, the pH value of the soil must be kept at the preference rating of the crops to be raised.

Limiting Factors

Too many farmers attribute poor yields to luck; the word is used to cover a multitude of sins and transgressions of vital natural laws or as an excuse for our own inefficiencies. However, in farming there is always the element of chance and probable error. Fluctuating yields are definitely caused by one or more limiting factors, and any condition that may limit plant growth may be regarded as a limiting factor in crop production. This may be a deficiency, absence, or unfavorable condition. It may be sunlight, tilth, moisture, soil reaction, soil aeration, organic matter, microorganisms, carbon dioxide, or any of the major or trace elements. When all factors are favorable, maximum yields result, but, if one is deficient, the yield will be reduced accordingly. Usually, low yields result from more than one limiting factor. Furthermore, what would constitute a limiting factor for one crop would not necessarily be a major limiting factor for another soil acidity, for example. Similarly, conditions causing outstanding yields on some farms or in certain fields might be almost entirely due to normal soil fertility limits in a given section.

A low-lying field to which special care had been given, for example, could be expected to yield up to 150 bushels of corn per acre due to an unfailing supply of underground water. It would, however, be poor economy to build up fertility on high ground with an eye towards such a bumper yield, as lack of moisture would surely diminish the crop. In short, each section of our farming area has its limits regarding moisture, and to try stretching this limit would only invoke the penalty involved in Liebig's law of the minimum. This law implies that nutrient elements are absorbed in certain definite proportions and in such manner that when the limit of one is reached absorption of the others is actually retarded or depressed.

This old law has been somewhat modified during the past decade, since it was found that in the case of the major elements—nitrogen, phosphorus, and potash—a deficiency of one of these elements results in an increased absorption of the other two.

Climate is another factor that must not be overlooked. If rainfall in a certain section of the Corn Belt is 13 inches during the growing season, this amount of water would justify a yield of 75 bushels per acre. It would, indeed be folly to add fertility sufficient to grow twice that amount for the fact that the law of diminishing returns would surely assert itself. It is a good practice to keep crop stands down to climatic levels, since the dry periods usually occur at some time during the growing season, and excessive plant numbers per acre only serve as an added drain on moisture reserves. Nitrogen has long been the bottleneck in crop production, but with increased knowledge in ways of trapping this element with legumes and improved understandings of the relationship between tilth and organic nitrogen formation, it is fast becoming a matter of secondary worry.

The most important factor in farming areas is moisture, especially the distribution of rainfall. The mean annual rainfall is quite sufficient for any crop if the moisture can be held in the soil as a reserve against drought. Nothing serves in this capacity so efficiently as a deep, oxygen-rich topsoil implemented by the furrow. This deep bed absorbs water freely and, with the aid of sponge-like organic material, holds a great amount of water against gravitational pull. Where the land is plowed at the same depth each year, there may be a tendency toward forming a plow sole that is impervious. This condition can be eliminated by plowing at different depths occasionally, so water has the benefit of the easier saturation in wet times and, in dry times, has more chance of being drawn up against gravitational pull.

There are some who claim that a deep bed falls easy victim to erosion; let's see. Suppose a violent rain fell amounting to 2 or 3 inches. Would not a deep, friable bed absorb more of this water than a shallow, tilled soil? The water that is not absorbed by the soil must necessarily run off, becoming the chief agent of erosion. The quantity of water that soils can retain against gravity is determined principally by their mechanical composition, the organic matter they contain, and the nature of their subsoil. The total quantity they may contain ranges from about 15 percent in coarse sand to about 134 percent in peat. The capacity of silt loams may vary from 30 to 45 percent. The approximate maximum quantity of available water that soils can contain varies from about 2.5 acre-inches to about 4 acre-inches. Since our chief concern is capillary water—water that is left in the soil and held against gravity after the free water drains off—it should be borne in mind that tilth and organic matter are chief aids in capillarity.

Moisture is the chief limiting factor in crop production, During dry periods, plants show definite hunger signs. We commonly term a yellowish cast to the leaves and stalks "firing," but, in reality, it might be a deficiency of nitrogen or some other soil nutrient. The soil minerals must unite with moisture to turn them into soil solutions to be available to the plant, and while the supply of critical soil minerals might be adequate under normal conditions, a long dry period would retard the action of these minerals. Bacterial action is also retarded in dry soil and may stop entirely under severe drought conditions.

The recent findings of experiment stations regarding the dust mulch controversy do not shake my faith in the least. During the seasons of natural moisture conditions, all that is necessary in the way of cultivation is to control weeds. A dust or earth mulch has no significance when there is a plentiful supply of moisture. Unfortunately, there are more dry seasons than normal or wet seasons, and during these dry seasons, the dust mulch comes to the rescue by holding back evaporation. To prove the point, it is only necessary to walk through a dry field and examine the ground. The soil in the affected field is invariably hard with the surface full of cracks. It appears much drier than the soil in less affected fields. Under the surface of a fine earth mulch, the soil is generally moist and conducive to the natural functioning of soil agencies.

Roots

Modern theories on tillage may influence human thinking, but not the roots of plants, which follow a logic all their own. Roots were intended to penetrate the soil, gathering soluble nutrients and anchoring the plant, holding it in an upright position enabling the leaves to receive maximum exposure to the sun. Where attention is centered on surface tillage, the plant roots have no inducement to penetrate the soil to depths beyond the supply of nutrients; lodging is the result. Most everyone has observed the habits of plants that grow around a straw or manure pile. The stalks grow rapidly and soon become top-heavy. On examination it will be noticed that the plant can be pulled up easily since the roots are shallow. The reason for this condition is simple: the roots find their requirement in moisture and nutrients close to the surface and go no further. This is largely the objection to applying fertilizer by the corn planter. The fertilizer attachment drops the fertilizer along each side of the row, and the corn roots, on contacting this fertilizer, have no inducement to go beyond the limit of their nutrient requirement, thus failing proper anchorage of the stalk. If there is insufficient moisture to dissolve the fertilizer so placed, the young roots may be injured by burning, limiting their growth. Crops that lodge do not mature fruit naturally, and their quality suffers along with their yields.

Lodging is not the only objection to shallow roots. Moisture naturally evaporates from the surface, and in long dry periods, shallow roots show deterioration first. A deep seedbed, full of sponge-like material with the nutrients worked throughout the soil, is the only satisfactory rootbed. The roots will feed through this soil at great depths and can perform their function in the natural way. If fertilizer is used, it should be put on with the moldboard plow and covered with a thick layer of soil.

Many are the ills following shallow tillage. If the soil nutrients were all placed in the top inch of the soil, the actual soil-building area would be greatly reduced. Dry periods would stop natural bacterial changes, and the depth of productive soil would be lessened. In spite of modern theories of tillage methods, roots will continue to do a better job of production if their natural habits are encouraged. Deep plowing fits into the natural scheme of plant nutrition and will grow and maintain a healthier plant.

As the guardian of soil fertility, roots have no equal. It is folly to allow erosion to rob our soil of nutrients that should go into the growing of crops. To the rescue has come Reed canary grass that seems to be dedicated to the task of holding the soil in the natural waterways. The flash floods of 1944 have shown the necessity of sodding the natural waterways on sloping land like mine. Contoured rows hold and spread normal amounts of water, but when the rainfall is excessive, the row ridges must dump the excess. This amount assumes great proportions if the slope is long and deep gullies are sure to result. Where a band of close-growing crops are used at regular intervals, the prospect of a gully forming is somewhat lessened, since the water will spread on coming in contact with the smooth, close-growing crop. The contoured rows, however, present a different problem, and a careful job of engineering must be done. The gullies formed on the contoured hillside will assume huge proportions during a single rain if the topography of the land has not been predetermined and sodded. If these gullies are allowed to develop, the whole soil-building program will have been defeated and the productivity of the soil lessened to a lower level than when the old up-and down-the-hill method of planting was used. Maintaining productivity is a comparatively simple matter if the soil nutrient loss from erosion and leaching is held to a minimum. It is easy to tell the amount of nutrients removed by crops, but it is difficult to determine the amount of productive power and soil structure lost by erosion.

To prevent erosion, a small patch of Reed canary grass be sown, and after a patch has matured, the sod can be cut up in small pieces and reset where needed. Single plants can be set 18 inches apart and, in just the first year, will thicken to a tight sod. This grass survives drought and also does well in wet places. When a gully or watercourse becomes sealed with Reed canary grass, it stays sealed. It stands silting extremely well, and when covered up with silt, has no trouble in piercing through. It is also valuable hay, and, if there is much to be sodded, it means that the land will not be entirely out of production.

As has been pointed out, soil fertility and productivity are a group of conditions any one of which, if missing, reduces the productivity and lasting qualities of the soil, so before labor and money are expended on a soil-building program, a prudent farmer will check the topography of his land. Maybe the field cannot hold its fertility without help.

When plowing up a sod field, I have often remembered the eloquent tribute paid this greatest of all crops by Senator J. J. Ingalls of Kansas and reprinted in *Your Farm* magazine. Senator Ingalls said in a few words what others might develop into a lengthy volume:

> Next in importance to the divine profusion of water, light, and air may be reckoned the universal beneficence of grass. Lying in the sunshine among the buttercups and dandelions of May, scarcely higher in intelligence than those minute tenants of that mimic wilderness, our earliest recollections are of grass.
>
> And when the fitful fever is ended, and the foolish wrangle of the forum and market is closed, grass heals over the scar which our descent into the bosom of the earth has made, and the carpet of the infant becomes the blanket of the dead.

> Grass is the forgiveness of nature—her constant benediction. Fields trampled in battle, saturated with blood, torn with the ruts of cannon, grow green again with grass, and carnage is forgotten.

While grass is worthy of this beautiful eulogy, other natural plants, commonly called weeds, should be given a fair trial in the court of agricultural fairness. Volumes have been written about weed control and what the intrusion of weeds in our cultivated crops cost the farmers each year in reduced yields.

Our soil of today is quite unlike what it was thousands of years ago, since countless generations of vegetation went into the forming of our topsoil. Without this lush growth of vegetation and soil-holding roots, there would have been nothing on this continent to have attracted the eye of the early explorers. Put simply, a nation is no better than its topsoil. Nature was a long time in forming it and, consequently, guards it zealously. Man's thoughtlessness is responsible for much of the devastating erosion and kindred evils that have become a threat to our national economy. While some weeds will compete with the cultivated crops for moisture and soil nutrients, they contribute their organic matter back to the soil, and their roots form a natural barrier against the ravages of erosion.

By contrast, when a crop is removed from the ground in the summer or fall, if this ground is free of weeds or natural grasses, this soil is vulnerable to the elements for many months, and an irrevocable damage is done. Instead, fall grasses should be encouraged to follow a cultivated crop where no provision is made for another cover crop.

Fall plowing is fast becoming recognized as the last remnant of predatory farming, for the removal of a sod covering far in advance of the planting date deserves no better term than predatory. Many defend fall plowing by pointing out that it makes the earth more amenable in the spring. Granted, it may look nice until the first heavy rain when the flour-like soil runs together and, on drying, becomes a hard mass. In the seasonal dry periods, fall plowing is the first to show curling and firing of the crop. Such soils pack quickly and do not allow a rapid saturation. On the contrary, spring plowing stays porous with enough coarse, earthy materials to keep it oxygenated. This condition is to the liking of our crop plants, since this aeration provides oxygen and nitrogen to the plant roots and soil organisms.

Soil will hold its own and grow more fertile if some sort of vegetation is allowed to remain—even weeds will do more good than harm for they are nature's own safeguard against erosion. I am by no means defending certain noxious weeds, but there are plants and grasses classified as weeds that work with nature in holding the soil where it was created and do a thankless job for the farmer. If we are not to abuse the use of the plow, we should not disturb the roots that hold the soil in place, until it is necessary.

All plant roots, not just legumes, have the properties of gathering or attracting nitrogen. The proof of this statement lies in the fact that bacterial population in immediate contact roots is much more dense than a fraction of an inch away. However, legumes are equipped with nodules on their roots to culture and hold great numbers of these organisms. For this reason they are the best field crop to rapidly replace nitrogen.

Inoculation of legumes has come into prominence as an aid to increase the bacterial population of the soil. We know that roots support living plants and supply them with water and soluble plant nutrients absorbed from the ground. All that is necessary to draw vital nitrogen from the inexhaustible supply of the air is to ensure an oxygen-permeated seedbed enabling free nitrogen to readily contact plant roots. This vital process cannot be accomplished properly in a shallow or hard seedbed, but, if nitrogen deposits are to keep ahead of withdrawals, it must be borne in mind that the only place nitrification can take place is in the topsoil. Plenty of nitrogen can be kept in the soil by natural means; this organic variety is by far the best, as there is no harmful reaction of the kind produced by a commercial product.

Soil Nutrition

While this work stresses the importance of organic fertilizers, there is a definite place for phosphorus and potash. And the rapidity with which they are used up by the growing crops makes it a vital necessity to look after the soil's need for these particular elements.

Phosphorus is by far the element that needs careful watching, since it is taken from the soil in large amounts and but little is replaced by our ordinary farmyard manures. We are spending this vital element faster than we think, and the time is here to do some actual figuring on the probable reserves. It is calculated that the best soils originally contained about 1500 pounds of phosphorus per topsoil-acre. On less well-balanced soil, the amount per acre may run as low as 500 pounds. If we consider that many crops remove as much as 14 pounds of phosphorus per year—and but a small percent of this is recovered and put back in the soil—we should consider it high time to look to other sources of supply. Grain farming results in a greater loss of this element, as the actual amount contained in the grain is lost from the soil on which the grain was raised. The rate of disappearance is not so marked with stock farming, since about 50 percent of the element is recovered.

Phosphorus has often been referred to as the master key to a permanent agriculture. This reputation has been gained, not because phosphorus is the most important soil element or because phosphatic fertilizers will produce larger yields than other nutrients, but because phosphorus occupies a somewhat unusual position in plant, animal, and soil relationship. Indeed, phosphorus is present in every living cell, and is closely related to the formation of new cells. In plants, this element is especially important in keeping up root development and seed formation. It hastens maturity and promotes a vigorous growth. In acid soils, phosphorus is not readily available, and, consequently, an application of lime will release that part that becomes locked up on account of the soil's acidity. If there is a need for phosphorus, an application of any standard super-phosphate will most certainly pay big dividends.

Farmyard manure may vary widely as to its fertilizer value, since the kind and management of the animals has much to do with its quality and use as a soil builder. The feed of animals reflects the value of the manure, and the state of decomposition determines largely its adaptability to the soil. Fattening animals utilize only 5 percent of the nitrogen contained in feeds used, whereas growing animals and milk cows utilize considerably more. Therefore, feedlot manure holds the greater value as a natural fertilizer, and the sooner it can be incorporated in the soil the greater the return.

It has long been supposed that it was unwise to apply lime and other commercial fertilizers together, as it was thought one element would reverse the action of the other. Careful tests have exploded this notion and give hope to those who like to compost their manure. This practice is gaining favor with many farmers and particularly truck gardeners who do not have sufficient animal numbers to utilize the roughage. To the waste roughage can be added lime and phosphate, that, when acted upon by moisture, are blended into an ideal fertilizer.

It is safe to assume that when a mineral element is found in the tissues of plants, there is a definite need for that element and, in case of its absence, that the plant cannot produce a profitable return. Very little credit has been given to trace elements, since only a very small amount is needed to supply the plant's needs. But a deficiency of one or more of these minor elements would affect the behavior of the particular plant involved as much as a deficiency of a major element.

Of the known elements occurring in nature, fifteen of the more important are found in plants: carbon, hydrogen, oxygen, nitrogen, sulfur, phosphorus, potassium, manganese, iron magnesium, calcium, sodium, silicon, chlorine, and aluminum. A very small amount of the less important elements are also found in most plants, and, though the amount is almost negligible, they no doubt play an important role in performance. They are boron, copper, zinc, iodine, arsenic, nickel, cobalt, barium, antimony, fluorine, lithium, titanium, strontium, tin, cesium, cerium, chromium, and vanadium.

Since hydrogen, nitrogen, oxygen, and carbon come from the air, we need never fear of a shortage of these elements if we set the proper trap and use the right kind of bait. The trap is a deep, oxygen-rich seedbed, and the bait is the legume.

Much is written about nitrogen and phosphorus, and very little space is given to the potash that works along those two as a third horse. Potash is a metallic element classed as an alkali. While it is never found free in nature, its salts are widely distributed. Some soils, especially clays, have an abundant supply of this element. It seems to be the chief residue of wood ashes. Lack of potash in a plant lessens its resistance to drought, causing the leaves to curl and fire easily. Potash seems to enable the stalk and leaves to hold more water.

By comparison, the lesser elements require a delicate touch in their application, as too much might prove toxic to the plant. But many areas are unproductive for the lack of these elements, and work is already underway to find ways of putting these soils in proper balance. In each case, plants show distinctive hunger signs when there are shortages of major or minor nutrients in the soil. Deficiency of nitrogen, for instance, is noticed by the unthrifty condition of the plant, stunted growth, and, in many cases, complete crop failure. Phosphorus tends to promote root growth and develop the plant rapidly, hastening the maturing of the grain; lack of phosphorus, on the other hand, causes stunted growth and late maturity. Potash, as has been said, works with phosphorus in storing moisture. Too much potash might cause the grain to be late in maturing; too little may cause firing.

While nitrogen, phosphorous, and potash are principal players in the soil drama, other nutrients play vital roles. Iron is essential in promoting the development of deep green leaf color, incorporating itself in the plant and grain in a form that is essential to both animals and humans. Magnesium is a constituent of limestone, and, on well-limed soil, there is little danger of a deficiency of this element.

Many soils are kept fertile by legumes and barnyard manures. The exception seems to be phosphorus, since barnyard manure is notoriously lacking in this element. It has long been noted that phosphorus is the first soil element needing replacement. Since it is an acid radical, it does not enter into base-exchange. It is easily fixed in the soil, but, on strongly acid soil, it is not so readily available to plants.

Beyond nitrogen, phosphorous, and potash, the cropper is advised to recognize other deficiencies at a glance. Magnesium deficiency is usually indicated by a yellowing of the leaves. While iron is needed in only very small amounts, a deficiency of this mineral seems to retard the action of the organs producing green chlorophyll. Sulfur is removed from the soil by leaching more rapidly than by cropping. However, the natural supply of this element in the soil is quite abundant and is replenished by natural means to near the yearly crop requirement. Manganese is present in most soils to meet the small requirement of plants, but in certain areas where it is deficient, serious effects have been noticed. Boron has a favorable effect on plants only when present in small quantities. Absence of this trace element results in crop failure; too much is also disastrous. Copper is necessary for plant growth and, like boron, it is needed in only trace amounts. Chlorine, while not as critical as the other, lesser elements, is found in most plants, and therefore must be considered a necessary counterpart.

Given the intimate connection between soil and grazer, we must not close this chapter without first discussing nutrients crucial to animal life. It has been found, for example, that no form of life could exist without manganese, and a gradual wearing out of this element would surely reflect on the vigor of the animal population in deficient areas. Cobalt, it has been discovered, tends to strengthen the blood against anemia. Zinc is of proven worth in animal nutrition, since it has beneficial effects on vitamin efficiency. Magnesium is essential in the animal diet if the animal responds to feeding, as it aids in fat utilization and strengthens teeth and supporting structures. Lastly, iodine, like many of the other trace elements, is of demonstrated value.

Soils lacking in any of these elements will not produce optimum crops, and the produce of such soils will not raise profitable animals. By feeding the soil these elements as they are needed, nutritional problems will be taken care of in nature's own way.

Base-exchange Simplified

When the base-exchange process is expressed by the chemist, it is phrased in symbols that are beyond the grasp of an ordinary farmer like me, but when the true facts are understood and expressed in the farmer's own style, it becomes an interesting story that nature intended us to know. The principles of base-exchange are embodied in the reasons for soil acidity. Lime is one of the elements that figure in base-exchange. A simple illustration of what happens to the calcium in the soil may be shown by the principle of the water softener. Calcium and magnesium in water make the water hard, and when this water passes through a solution containing sodium, the calcium and magnesium are absorbed while the sodium is set free and passes into solution. Common salt will also absorb calcium and is the natural enemy of other soil agencies affecting soil fertility.

Because calcium is easily absorbed by sodium and other salts usually contained in commercial fertilizer, the reason for the natural hangover following the use of nitrogenous fertilizers becomes clear, especially since the nitrogen content is usually supplied by sodium nitrate or ammonium sulfate. Usually, only a limited quantity of an exchangeable element of a soil material can be brought into solution by the single addition of a displacing agent. But under leaching conditions, the displacement can finally be made complete. Most soils have some absorbent element as a natural heritage and affect such base elements as calcium, magnesium, potassium, and sodium, but the elements of acid radicals such as nitrogen, phosphorus, and sulfur are not affected. When nitrate, sulfate, and chloride are passed through soil materials, the basic elements are absorbed while most of the acid parts come through as filtrates.

Base-exchange is recognized as the most important fundamental principle of soils. In it are found explanations for several important soil phenomena or conditions, including soil acidity and alkalinity, friability of some clays, fixation of potassium and ammonium, non-fixation of nitrate nitrogen, and tight or impervious clays.

Base-exchange involving both organic and inorganic soil colloids is only one of the important principles concerned in soil fertility. It should be borne in mind that other important chemical reactions take place in the soil. Not all soil constituents take part in base-exchange, although most of them are affected in some way by the process.

It has been a common belief that upland clay soil was less acid than lower-lying land, but the facts suggest the contrary. In humid climates, this upland soil has been rendered acidic by leaching, and, in many cases, the lowlands have caught and held much of the basic elements that at one time were in the higher soil.

When a potash fertilizer is applied to a soil, the potassium is soon fixed—which is to say it is absorbed by base-exchange materials. Sometimes considerable potassium may be absorbed, often with the release of equivalent quantities of calcium and magnesium. Thus, potash added to the soil may not increase the amount of potassium in the soil solution to a great degree, but instead it may set free a greater amount of calcium magnesium. The fixed potassium becomes a reserve for plants.

Nitrate nitrogen is not easily fixed in the soil, as there is no agent for this purpose; however, ammonium nitrogen is temporarily fixed. It seems that organic nitrogen is natural for fixation in the soil and becomes a more dependable source.

Since phosphorus does not enter into base-exchange, it is easily fixed in the soil, and a good thing, too, for phosphorus is by far the most important soil mineral nutrient affecting soil fertility.

Alkali soil presents a different problem, but such soils are being reclaimed and made productive. Alkali soils seem to have a superabundance of soluble salts, sodium, and calcium but are in such an unbalanced condition that the constituents are locked. In some cases the sodium can be leached out by flooding and draining, but, where this is impossible, it may be advisable to use sulfur, which will bring the calcium into solution and neutralize the carbon dioxide.

And so the principle of base-exchange is one of the miracles of nature, without which there would be no fixation of critical soil nutrient minerals to be fed to the plant as needed. As can be determined from the aforementioned processes, if the principles included in base-exchange are kept in balance with the acid radicals, the soil can be said to be fertile.

But lest we rest complacent in apparent understanding, it should be mentioned that there is still much to be learned about this vital process of nature and its linkage to the health and vigor of plants. A broader knowledge of its effect on the lesser elements or trace elements might reveal interesting details that will add a chapter to life's health story. The importance of moisture as a soil-building or soil-depleting agent is well established. Excessive moisture not only speeds depletion by rapid leaching, but aids acid-forming agents in intensifying the base-exchange of the major element. The chief concern, at present, should center on the fact that soil fertility is a matter of organic and inorganic materials, each useless without the other.

That nitrates and sulfates, as nitrogen carriers, have no value in permanent soil-building programs has been proven by the fact that nature has no way of fixing them in the soil and also on account of their acid tendencies. The fact that phosphorus is easily fixed makes super-phosphate the real payoff element.

Fertility

There is much confusion in the general understanding of the words *fertile* and *fertility*, and many believe that fertility may actually be leached from the soil. The only difference is in the words themselves. Fertility is a noun and fertile is an adjective. As a matter of fact, fertility cannot be leached from the soil, as it is not a part of the soil but is, instead, a quality of it.

When the soil fertility is removed by leaching, nutrients are lost and the soil changes from fertile to infertile. Therefore, fertility and infertility may both be regarded as relative soil qualities. Fertility is a certain degree of soil productivity, below which the term has no logical meaning. A soil may be fertile and productive for a certain crop but infertile for another. A soil may be ideal for a bumper crop of corn or other grains and yet infertile for alfalfa or sweet clover.

Many factors enter into soil fertility, obviously: good tilth, moisture, aeration, favorable soil reaction, organic matter, microorganisms, nutrient elements, and rotation. A soil may produce a low yield of corn for the want of phosphorus, but this does not imply that the yield can be increased in proportion to all amounts of phosphate that may be added. Increase in the amount of one element added to the soil may increase yield to a certain point at which further yields might be limited by the lack of some other element. In reality, lack of moisture in certain years may render a fertile soil infertile. Soil fertility, therefore, is determined by several, named factors whose combined effects are expressed in the ability of the soil to meet specific crop needs.

A fertile Florida soil, producing luxurious native crops, would be an infertile Iowa corn soil, and so on. In a general way, it may be said that, for most crops, the state of the soil in regard to its natural or acquired productivity has a much greater effect on yield and quality of product than have the characteristics that determine soil types.

The real holy grail of soil fertility is to find out how to meet the needs of plants through soils in the best manner, as may be determined by plant requirements, farm economy, and the effect produced upon the soils and plants by methods implemented. Where other conditions are favorable, sustained or increased yields result when soils fully meet plant needs. Fertility is, therefore, the balanced condition of the soil constituents and cannot be brought about by the addition of any one of these constituents if more is required to meet this balance. In many cases, lime is the determining factor in fertility, although lime itself is not classed as a fertilizer; it alters the character of the other elements through base-exchange and encourages their function as soil nutrients. Soil may be classed as fertile, and yet, by its physical character, it will not produce profitable crops. Compact soil, to name one inhibiter, will not permit aeration, without which soil agencies cannot function and plants will not thrive. Wet soil has much the same character with respect to aeration, coupled with the toxic effect that prolonged wet conditions provoke. Long dry periods, while not affecting aeration, retard bacterial action and deprive the plant of soil nutrients that must be made soluble.

To return to my thesis, if the soil were plowed at different depths, the organic matter would be distributed more evenly through the soil. It would be beneficial to have a supply of organic matter at lower levels, thus encouraging soil organisms to draw nitrogen deeper into the soil. This method would accordingly induce the roots of plants to further penetrate the soil, thus providing greater support. Deeper plowing would also increase the water-holding capacity of the soil, and, with a goodly amount of humus to act as sponges, firing would be delayed in dry periods. This furrow would have a still a greater advantage since the water from ordinary rains would soak into the ground instead of running off from the shallow, plowed ground, causing unnecessary leaching away of vital soil nutrients.

There is a close relationship between fertility and temperature, though this fact has been generally overlooked. Bacteria that cause decay act less rapidly in cold soil and stop entirely at freezing temperatures; consequently, shorter seasons, such as those in the more northern climes of North America, would not use up the current crop residue, and this growth accumulates in the form of moss. The farther north, the thicker this moss becomes. Even in the northern part of the grain belt, organic matter will not wear from the soil as rapidly as in the southern part. In the Deep South, where the ground never freezes, bacteria are constantly working and fertility can be accomplished more rapidly, but soil depletion is accordingly more rapid.

Present interest in contouring rolling ground will do much in preserving our soil, but the most important problem at present is to find a quick-growing, fibrous, shallow-rooted grass to serve as a cover crop to halt leaching, for this leaching represents the most pressing threat to our healthful existence.

The Value of Rotation

It is a well-established fact among farmers that any particular crop does not follow itself without a decided shrink in yield. There are, of course, many scientific reasons to substantiate this claim. Carefully kept plots, covering a long period of years, have proven that a rotation of crops such as clover, corn, and oats maintains fertility and yields to the same levels as have manures and fertilizers. In truth, the economy of keeping productivity at high levels by simply rotating the crops might well spell the difference between success and failure.

Soil sanitation is a vital subject and is rarely discussed. Most of the diseases and insects that are injurious to crop plants are markedly limited by the number of species on which the organisms concerned can live. Many such insects and diseases are destructive to only one kind of crop. Continued growing of one kind of crop on the same land favors the development of these insects and diseases until, ultimately, the crop plants are unable to resist their attacks. Growing crops that are not attacked by some such insect or disease organism results in their destruction, because by so doing, the life cycle of the injurious organisms is broken.

Continued growing of cultivated crops results in a most rapid destruction of soil organic matter. In crop rotation, on the other hand, destruction of soil organic matter does not take place so rapidly. It has been warned that bare soils lose large quantities of nitrogen as nitrates, and that this loss may be greatly reduced by growing crops. Commonly, soil nitrates are allowed to accumulate and suffer loss by leaching during the late summer after some crops have been harvested and when soils are often left bare. If the natural fall grasses, commonly called weeds, do not appear, it would be advantageous to plant some cover crop that could be used as pasture while it holds down the natural loss from erosion.

The value of rotation deserves more credit. I have formed many conclusions through long years of farming, and, despite what is being said of fertilizers and all such plant stimulants, I am steadfast in my belief that nature holds the secret in the production of a big, sound ear of corn, and if and when this secret is learned, it will have to do with rotation. The natural nitrogen-fixing agencies have no serious competitors, and to me it would seem that chemical nitrogen or commercial nitrogen has no place in any soil-building program.

In addition to the beneficial effects on soils and soil fertility that result from alternating crops, rotation affords opportunities for improving soils in other ways and for realizing greater returns. For example, agricultural lime can be applied to the most responsive crops and at such time as to effect the best results with all crops in the farmer's rotation. Greater immediate returns from the use of lime are commonly realized when it is applied for crops like clover and alfalfa. The soil changes brought about by the added lime and the increase in soil organic matter resulting from the better growth of clover invariably produce greater effects on succeeding crops than if lime were applied directly for the other crops. The same reasoning holds true with the commercial fertilizers that, when applied to grass crops, induce a greater vegetable growth that, in turn, adds fertility and beneficial physical properties to the soil.

We are beginning to realize that plant roots excrete certain substances that are utilized as nutrients to plants of different species, and if the crop plants are not rotated, these excrements build up toxic acids that make continued growing of that crop unprofitable. On the contrary, a change in species may utilize these excrements as a nutrient, and a profitable crop results.

Different plant species may be compatible or incompatible. For example, timothy and red clover may grow side by side apparently with mutual advantage. The same is true with many other field and truck crops. Natural plant associations also suggest such beneficial or mutual relationships. It is suggested, for instance, that strong foragers for mineral nutrients should not follow plants that have removed the same kind and amounts of these nutrients, but rather a plant that uses a smaller amount of these nutrients should follow the stronger foragers.

It should be borne in mind that deficient soil limits the ability of the plant to act as the carrier of certain mineral nutrients, and only by careful handling of the soil can we keep the forces of life in our food plants in their proper balance. Should a farmer, for instance, buy corn for livestock feeding and find this corn was the product of a soil devoid of the necessary mineral nutrients, it necessarily follows that the corn would be of a lower protein content, thus having a lower feeding value. It is noteworthy to consider the difference in protein content of corn from different soils. At present there are recorded tests that show a variation of from 4 percent protein to 11 percent protein in corn from poor soils and good soils respectively. This knowledge becomes of vast economic importance to livestock feeders everywhere, as it explains why some cattle and hogs do so much better and put on more economical gains than others. The situation is just as grave in our truck and garden crops. For example, the carrot is praised for its heavy inclusion of minerals and vitamins, but, if raised on soils that are not balanced with the necessary mineral nutrients, it can hardly be expected to be a healthful carrier of these same elements.

For many thousand years, all seeds sown in the fields were broadcast by hand. This practice did not permit subsequent tillage. Flemish farmers found the advantage of row cropping, and this method was copied by western agriculture. The fact that it took many thousands of years to work out the underlying principles of crop rotation illustrates how difficult it is sometimes, in spite of years of practical experience and creative thinking, to make simple, crude discovery a scientific practice.

PART THREE

Soil & Sentiment

Chapter XI.

SOIL AND SENTIMENT[1]

No man has really fathomed the mystery of growing food from soil. Few realize the magnificence of the processes of plowing, planting, sprouting, growing, and harvesting. Still fewer realize they are working in cooperation with God.

Owing to this high calling, we farmers are placed in a strategic position of trust and responsibility; we must put back in the soil what is taken out. We are not miners, but husbandmen of the soil, so that the soil will continue to be source of life and pleasure for generations to come.

Some have been mining the soil with reckless abandon, and for them the day of reckoning will surely come. Since too many have taken the gift of the soil for granted with little heed for the future, millions of acres of good land have become infertile, and many millions of acres ought to be retired from cultivation without delay. This frightful waste of the soil was brought about largely by people who did not anticipate the future and paid little attention to crop rotation and erosion, but instead merely planted crops for cash each year. We are not paying back the wealth we take from the land, and, as the land washes out, blows away, or loses its productivity through improper cropping, our rural population becomes less and less in numbers and efficiency. Since permanent agriculture and permanent soil fertility are one and the same thing, our national strength depends on our willingness to keep our soil fertility permanent and to reclaim those millions of acres that have been ruthlessly exploited.

Because soil is always in the making, it must be remembered that the soil we really lose and need worry about is that which is washed away. Erosion carries the soil from the farm and deposits it at the mouth of some river. Such soil ultimately becomes of some use, but the land that it leaves is rendered infertile for years to come. If land is reasonably level and cared for so as to stop erosion, there is no such thing as wholly exhausting its fertility, for each year some new soil is created. Nature has a way of protecting itself, but good farming requires that the land be fed. Although soil tests, formulas, and rules for maintaining productivity of our soil are useful and necessary, what we need first is a widespread philosophy of permanent soil fertility for keeping up the soil now and for future generations. This urgency responsibility arises from a fundamental love for the soil.

Many people do not comprehend the many changes that are constantly going on in the soil—biologically, chemically and physically—and these changes determine to a large degree the quantity and quality of the crops the farmer removes from the ground. Nor do they understand what, to all peoples of the earth, good land means: life, success, and well-being. Who can plant a seed in the ground, watch it sprout and grow and reproduce in abundance to feed a hungry world and not love the soil that mothered the seed? Who can walk barefoot in the freshly turned furrow as a youngster and not feel the thrill of it? Who can crumble the rich, black earth in his hands and not realize the preciousness of it to human life nor feel excitement at all the mysteries it holds? Who fully realizes the high calling of farming and the possibilities it holds mentally and financially and not love the soil that makes this good living possible.

The well-rounded farmer is not a wage-slave but a creative worker. A good farmer who loves his soil and realizes its powerful meaning to humanity takes pride at the end of the year in having his farm a little more productive, his home a little more beautiful than when the year began. Within this ideal lies the possibility of permanent soil fertility. Because a man owns a farm does not mean that he owns the land to do with as he pleases, for, in reality, he is only its tenant. He has use of the land for his lifetime, at the end of which he should leave the land to future generations in as good or even better condition than he found it. We may not waste that which is not ours.

Every farmer has a divine obligation, a sacred trust, to preserve the productivity of the soil, and we must not deprive future generations of good soil or understanding of the earth's goodness. The pride one feels as they walk through the fields on a spring day is truly rooted deeply in nature. The smell of spring awakens new interests in the tasks that lie ahead. The lilt of the meadowlark serenades the denizens of fields and woods.

It is spring, Easter time—fragrance, color, music, hope. While one stands there amid all this, the emotional upsurge becomes ageless in the heart. Many things thought important at the time are swallowed up by the swiftly moving events, but there is always something—perhaps a single incident or a particularly pleasant day or maybe a song—that will cling to us throughout our lives. I have felt this many times, and my emotions seem to center on a particular Easter Sunday many years ago. A pleasant south wind was awakening new life under the winter-whipped stems of the last season's vegetation. My daughter Helen emerged from the house bedecked in her

Easter finery, humming that grand old song, Easter Parade. Her sentiment to me seemed singularly appropriate, for in it was the spirit of Easter, as far as her juvenile mind could reckon. As she backed the car from the garage and turned toward town to join the Easter Parade, I did not realize, at the time, that this day embodied life's force and its greatest attribute. Neither time nor eternity can take away the particularity of this day, and, as time passes, the charm of it remains radiant, immortal. To me the greatest force is the music of spring and the planting season.

For every person there is a spot, a place they pledge themselves to return some day, because they were happy and contented there. It, no doubt, is the old home, the haunts of their childhood. These places are of the earth and the earth's magnetic power. Some get back after many years and are greeted by the same landscape, just as they left it—the same old hills and streams and shrubs untouched by anything save the changing seasons.

My place is under the Iowa skies, floored with the world's best soil and walled in by the emerald skyline of native oaks. I have planted and harvested these fertile fields for many years and have learned from mother nature that she holds no grudge. She has healed the wounds and scars I thoughtlessly made on her contour and patiently replaced the fertility I so lavishly spent. No, this is not agricultural science, but it does have to do with the story of the soil I have set out to tell, for the lack of sentiment in our associations with nature means the lack of interest in the business of farming.

My sentiments do not draw their substance from my imagination; I have lived them; they are real and everlasting. And still today, the same reverie comes each spring; the scenes are the same—everything, in fact, save for the little girl. She has grown up now and has gone, but imagination keeps her on the set, and her role will always be that of the leading lady in the drama of spring. In thinking of her I realize greed often shuts our eyes to the really good things in life and hardens our hearts to the many little things that would add up to a more pleasant and useful life.

As I struggle to collect my thoughts and equilibrium in the morning after a fitful night, what a lift it gives me to be greeted at the back door by my dog Simon. He goes through with his antics; he is relaxed (only animals know how to relax), and is ready to follow at my heels wherever I go. This demonstration of affection does more for a man than the morning coffee. Good old Simon! He has no need for tobacco or liquor to be congenial. If I tell him to wait for me, he waits, ignoring personal comfort; all that matters is my return.[2]

When the going is difficult, the weather drab, and you can but look out over the fields resting under their thick blanket of snow, time moves slowly. But in lighter moments you are heartened by the assurance that spring is coming; there will always be spring; there will always be Easter with its new life and promises.

What has all this sentiment got in common with tillage methods? Just this: new and abundant life cannot emerge from an impoverished soil but must be the result of diligence and due regard for the future. The degree of joy, over and above bare necessity, can be realized only to the extent that we have planned for it. If we want to enjoy the mirth and gaiety that are necessary in building a lasting culture, we must conclude that such energy can only emanate from people who are well-nourished.[3]

Appendix

FROM THE JOURNALS OF WALTER THOMAS JACK

The following fragments were selected from the farm journals of Walter Thomas Jack for their insight into the literary, impressionistic, and highly moral nature of Jack's agrarian reflection. They are presented here in no particular order and with minimal editing.

- Waiting for the thresher, the shocks of grain stand with tussled heads, like small boys who have forgotten to comb their hair.

- The small-waisted, wide-skinned corn shocks rustle in their rows across the field, as though any minute they will move off in an old-fashioned dance.

- A single cardinal sat singing in the tops of a blooming crabapple tree—a bright candle burning on spring's snowy, perfumed alter.

- The corn, gaunt but resolute, stands in the frosty dawn.

- He saw a hundred Hereford calves taking their morning nap beside the waterhole—and each calf wore that smug expression that only the pure-bred rich can achieve.

- The dry corn blades rattle like a family skeleton, and in every corner of the yard, the maple leaves are congregating to gossip.

- A sunflower patch had seeded itself along the fence, and one flower, in growing, thrust its head between the palings; then, having dried there, it gave the morbid appearance of having deliberately hanged itself.

- Product of the times.

- In my henhouse, the individual must be sacrificed to the good of the whole. I am degraded by this practice, but I fall in line just the same. Some day I may revolt; some day, instead of destroying the sick hen lest she infect the others, I shall destroy the others and nurse the sick back to glowing health, thus reestablishing my self-respect.

- In a voice like a cold front.

- The living mingle with the dead.

- Not until after economic security followed the Depression, did more lush times afford a virtual semi-cure, and he was able to concentrate on writing.

- Indeed, I am indebted to my contemporaries, although my work is extremely original. Only those who do not comprehend the enormous tradition of indebtedness between writers for the stream of literature can think otherwise.

- I do not think of myself as a writer. I am a farmer at heart, with literature as a hobby.

- Toward more picturesque speech: I walked along the placid lake—frogs trilled—silent as I came—loud again as I passed by.

- I walked through the meadow on a late-summer day. Crickets shrilled in the tall grass and flowering weeds.

- He was flicking pages in the book of his life.

Notes

INTRODUCTION

1. David B. Danbom, *Born in the Country: A History of Rural America* (Baltimore: Johns Hopkins University Press, 1995).

2. The page numbers quoted here and elsewhere in the introduction belong to the original edition, though the quoted language, edited for clarity, is drawn from the updated version.

3. Background information on Herbert Hoover and his family is derived from *World Book Multimedia Edition*, 2003 edition, version 7.1.1. World Book, Inc., 2002.

CHAPTER I: THE OLD QUAKER'S NATURE: A PREFACE

1. To avoid duplication with the new editor's preface, the original "Author's Preface" has become Chapter I, and it has been augmented with material concerning the scope of *The Furrow and Us* excised from Chapter VIII, "Field Vs. Plot," as it appeared in the original edition.

2. The "most ardent critic of the furrow" references, unequivocally, Edward Faulkner, though modesty prevents Jack from naming Faulkner in the original text. Dorrance promotional materials explicitly name Faulkner as the object of Jack's rebuttal, hence his name is introduced here and elsewhere as suggested by context.

3. This sentence marks the beginning of paragraphs incorporated from "Field Vs. Plot." A new transition proved necessary here.

4. This sentence marks the return to the original sequencing of the "Author's Preface."

CHAPTER II: THE FORCE OF CIRCUMSTANCE

1. The original chapter of the same name begins: "The facts and conclusions set forth in this work did not come easily but were born of necessity, the product of the tedious trial and error method that is unconsciously practiced by the tillers of the soil."

CHAPTER III: IN THE BEGINNING

1. "In the Beginning" was numbered Chapter I in the original edition.

2. The original chapter of the same name begins: "That our earth was created is unmistakable. That natural forces were geared to meet the needs of an ever increasing population is also something easily proven."

CHAPTER IV: A SCIENCE IS BORN

1. "A Science Is Born" was numbered Chapter III in the original edition.

2. The original chapter of the same name begins: "Primitive people could build only temporary dwellings, since after a few years of raising their food on the same ground, the diminishing yield would not feed them and they were forced to move to virgin land." The current opening picks up the narrative in the original fourth paragraph and proceeds chronologically.

CHAPTER V: WHY WE PLOW

1. "Why We Plow" was numbered Chapter IV in the original edition.

2. Now called Iowa State University of Science and Technology.

CHAPTER VI: NATURAL ECONOMIES

1. In the original edition, this chapter was titled "Natural Processes" and was numbered Chapter V. The title has been slightly altered to better reflect Jack's emphasis on the business of farming.

CHAPTER VII: MAN AND NATURALIST

1. In the original edition, this chapter was titled "Man or Monkey?" and was numbered Chapter XVIII. The title has been slightly altered to better reflect Jack's emphasis on Darwin and the author's preoccupation with soil-centric evolution.

CHAPTER VIII: HIDDEN HUNGERS

1. In the original edition, "Hidden Hungers" was numbered Chapter XIX. It has been moved to Part One of this new edition to reflect its essayistic qualities.

CHAPTER IX: THE OLD QUAKER WAS RIGHT

1. In the original edition, "The Old Quaker Was Right" was numbered Chapter VI. It has been moved to Part One of this edition as a logical bookend to the Author's Preface and a fitting conclusion to the first part of the new tripartite structure.

CHAPTER X: SOIL AND TILTH: A PRIMER

1. In the original edition, "Soil and Tilth" was numbered Chapter IX. It is employed herein, in combination with a new subtitle, as the umbrella title of a single section combining the more technical chapters (Chapters IX through XVII) in one comprehensive soil manual. This reorganization is in keeping with the original intent of the book, which was marketed as a "dramatic report of a dirt farmer" as well as a "helpful reference." The new A-level subheads necessitated by this reorganization correspond exactly to the original chapter titles, with the exception of "Roots," which was originally titled "Root-ology" and numbered Chapter XIII.

CHAPTER XI: SOIL AND SENTIMENT

1. In the original edition, "Soil and Sentiment" was numbered Chapter VII. It has been moved to Part Three of this edition to demonstrate Jack's union of the scientific model with subjective agricultural narrative.

2. At this point in the original edition, Jack included a brief poem in praise of his dog, Simon. The poem, prefaced by the disclaimer "A real Poet could do well with the following thought" read: "If I should die in some lonely place/Where friends they never knew/I'll bet a buck against your ace/Old Sime would die there too." The poem was moved to the Notes section of this volume because, in the editor's view, it tips the scales too far towards "sentiment" in what the new organizational strategy argues is a chapter evidencing a union of objectivity and subjectivity.

3. The current conclusion of "Soil and Sentiment" reflects deletion of the last sentence of the original chapter, which read: "Keeping the Soil productive above subsistence levels will insure for us the continuation of our national living standards." The intent here was to end on an appropriately lyrical rather than merely propagandistic or patriotic note.

About the Editor

Zachary Michael Jack, a fourth generation farmer's son, was raised on the Iowa land his family settled prior to the Civil War. Jack is also the editor of *Black Earth and Ivory Tower: New American Essays from Farm and Classroom* (South Carolina University Press, 2005) as well as the author of a limited edition chapbook of prose poems and parables entitled *The Story of Grief* (Oeoco Press, 2000).

The recipient of the Prentice Hall Prize for Poetry and the founding director of the Iowa School of Lost Arts for children, Jack has been a writer-in-residence at New York's Blue Mountain Center, Mexico's Great River Arts Institute, and Ireland's Tyrone Guthrie Centre. Currently, Jack is an assistant professor of English at North Central College in Naperville, Illinois.

www.ingramcontent.com/pod-product-compliance
Lightning Source LLC
LaVergne TN
LVHW022117200926
848481LV00003B/1186
* 9 7 8 0 7 6 1 8 3 3 1 7 8 *